Saul David is the author of numerous books, including *All the King's Men* and the *Zulu Hart* novels. He is professor of War Studies at Buckingham Univeristy and also a TV presenter.

Also by Saul David:

The Homicidal Earl: Lord Cardigan
Mutiny at Salerno: An Injustice Exposed
Churchill's Sacrifice of the Highland Division: France 1940
The Indian Mutiny
Zulu
Zulu Hart
Victoria's Wars
The Rise of Empire
All the King's Men: the British Soldier from the Restoration to Waterloo
Hart of Empire

Military Blunders

Saul David

Constable & Robinson Ltd
55–56 Russell Square
London WC1B 4HP
www.constablerobinson.com

First published in the UK by Robinson,
an imprint of Constable & Robinson Ltd, 1997

This revised edition published by Constable, an imprint of Constable & Robinson Ltd, 2012

A copy of the British Library Cataloguing in Publication Data is available from the British Library

ISBN: 978-1-78033-493-6 (paperback)
ISBN: 978-1-78033-861-3 (ebook)

Printed and bound in the UK

1 3 5 7 9 10 8 6 4 2

To Tristan

Contents

Maps

Preface

'Forward the Light Brigade!'
Was there a man dismay'd?
Not tho' the soldier knew
Some one had blunder'd
Their's not to make reply,
Their's not to reason why,
Their's but to do and die.'

Thus reads the best-known verse of Alfred Lord Tennyson's famous poem *The Charge of the Light Brigade*, written in the aftermath of one of the most notorious blunders in military history when a brigade of British light cavalry was sent into a valley bounded on three sides by Russian soldiers and cannon. Of the 676 riders who charged, only 195 were still in the saddle when the action was over. Their fallen comrades had paid the price, Tennyson believed, for errors made by their commanders.

Tennyson was right to pin the blame on Lords Raglan and Lucan for that disastrous charge in 1854. But military disasters are not caused solely by incompetent field commanders: inadequate planning (the Dieppe Raid), interfering politicians (Hitler at Stalingrad), overconfidence among officers and men (Custer's Last Stand), and even a poor fighting performance by the soldiers themselves (the Americans at Kasserine Pass) are all contributory factors. Most catastrophes are caused by multiple blunders – not to mention the talents and tenacity of the opposition – but one tends to stand supreme. For the sake of simplicity, therefore, I have divided the book into five themed chapters, each containing six case studies of disasters with a similar cause. Some are well known; others less so. But they all have one thing in common: they were largely self-inflicted.

Much has been made of the poor martial performance of certain nations, particularly in modern times. Contrary to popular belief, however, Italy does not have a monopoly when it comes to battlefield ineptitude. If this books proves anything, it is that no power, however vaunted its reputation, is immune from military catastrophe. The Roman Empire, medieval France, imperial Britain, the United States, Spain, Germany and Israel (see Afterword) have all suffered the ignominy of avoidable defeat.

Since this book was first published in 1997, it has gone into countless reprints and been translated into languages as diverse as German, Hungarian, Polish, Swedish and South Korean. The latter publication was the result of an email from a Colonel Park of the Republic of Korea (ROK) Army, a former commander of the Seoul University Officers Training Corps (OTC), who said the German edition had made a 'great impression' on him, and that a Korean translation would 'contribute to the improvement in the ROK army's combat ability'. Naturally I was flattered. But only up to a point: the purpose of the book was never to prepare armies for war; rather to warn politicians of the many pitfalls if they do take that fatal step.

The biggest sales of this book outside Britain and the United States have been, oddly enough, in Germany. A few years ago, during an interview for German TV, I was asked why a country that had turned its back on war was so fascinated by military blunders. I replied, somewhat mischievously, that it must have something to do with the fact that German troops were invariably the opposition. It was a throwaway line, but strangely true: of the 30 case studies, a third involve 'German' opposition (if we extend the term to include the German tribes of the 1st Century AD, Prussia in the 19th Century and the Austro-German troops that fought at Caporetto in 1917). The only other serial opponents in the case studies were the Boers of South Africa (three times), the Russians, the Turks and the English/British (all twice). The selection is, of course, arbitrary and heavily weighted towards British military disasters (there are 16 in total, not including the English at Bannockburn); yet the point holds good that the quality of the opposition is a factor in many serious military setbacks.

Can we learn from history? Certainly, though modern politicians seem stubbornly oblivious to the lessons of the past. Take Afghanistan. The unprovoked and ultimately disastrous British invasion of Afghanistan in 1839 was undertaken, like the 2003 incursion into Iraq, with regime change in mind: to replace a seemingly anti-British and pro-Russian ruler, Dost Mohamed, with a pro-British one, Shah Shuja. There, too, the plan was to withdraw British bayonets as soon as the country was pacified. It never happened, and tens of thousands of British, Indian and Afghan lives were lost in the ensuing three years of conventional and guerilla war. The end result: British troops finally withdrew, having first blown-up Kabul's magnificent covered bazaar, and Dost Mohamed resumed his rule.

Yet the lesson was not heeded, and three times since Afghanistan has been invaded by foreign troops: twice by the British and once by the Russians. Now we are back again, ostensibly at the request of a pro-Western government Kabul government trying to find its feet. And once again, as in Iraq, the very presence of foreign troops is making the security situation worse.

In Iraq, too, there was ample warning from history. It was

Britain, after all, which effectively created modern Iraq when it demanded a mandate over the former Ottoman provinces of Basra, Baghdad and Mosul in the aftermath of the First World War. This was partly because of Iraq's strategic importance at the head of the Persian Gulf, but chiefly because of oil: huge reserves had been discovered in both Iraq and Persian (modern Iran).

Within months, angry at the imposition of British rule, the Iraqis rebelled in Mosul and along the Euphrates. Railways were cut, towns besieged and British soldiers murdered. The British reacted harshly, dispatching punitive expeditions to burn villages and exact fines. They also used planes to bomb and strafe strongholds. By the end of 1920 a shaky peace had been restored, and by mid-1921 the throne of Iraq had been offered to Emir Feisal, son of the Sharif of Mecca, who had fought with Lawrence of Arabia. But Feisal proved less pliant than Britain had hoped, and in 1932 Iraq joined the League of Nations as an independent state. In 1958, Feisal's grandson was ousted in a coup that established a republic. And there Britain's interference in the internal affairs of Iraq came to an end.

Until, that is, the 2003 invasion. Many have argued that the US and Britain missed a golden opportunity to oust Saddam Hussein in 1991. In truth, the decision not to march on Baghdad after the liberation of Kuwait was not only considered but correct. 'We would have been there in a day and a half,' wrote General Sir Peter de la Billière, the British commander. 'But in pressing on to the Iraqi capital we would have moved outside the remit of the United Nations' authority, within which we had worked so far. We would have split the Coalition physically, since the Islamic forces would not have come with us. The whole of Desert Storm would have been seen purely as an operation to further Western interests in the Middle East.'

There was also a realization that toppling Saddam was one thing, replacing him with a stable, pro-Western regime quite another. 'If our soldiers depose him, or our special forces assassinate him,' wrote the then US Assistant Secretary of State, John Kelly, 'we risk losing American lives, bringing chaos and revolution to the region, jeopardizing the oil and, after all, his successor could be even worse.' Nothing much had changed by 2003.

The most recent military blunder in the original version of this book was the ill-fated Bravo Two Zero operation during the first Gulf War (1991). To bring the story up to date, I have included an Afterword with examples of blunders from the wars in Somalia (1993), Sierra Leone (2000), Iraq (2005) and the Lebanon (2006). All these conflicts were asymmetric (or low-intensity), pitting conventional forces against an unconventional opponent. A debate still rages in the higher reaches of the United States and British militaries as to the most effective way to defeat irregular forces: counter-insurgency (using regular forces) or counter-terrorism (using special forces). In the absence of a clear consensus one way or the other, they continue to use both methods and 'victory' (if that is possible in such a conflict) is as elusive as ever.

Only one thing is certain: as long as wars are fought, mistakes will be made. The trick, if at all possible, is not to fight in the first place.

Chapter 1

Unfit to Command

Incompetent generals seem to have a number of characteristics in common. Advanced age and physical incapacity are two, although General McClellan, one of the subjects of this chapter, was just thirty-five and in robust health when he commanded Union troops at Antietam.

An even more typical handicap seems to be a lack of command experience during wartime. Of the six generals featured, only one, McClellan (again!), had previously commanded in action a formation larger than a battalion – and then with decidedly mixed results. The others had either spent much of their careers on the staff or had recently been brought back from retirement – or both. Such battlefield naïvety tends to promote vacillation and overcaution, resulting in lost opportunities and, ultimately, disaster.

Elphey Bey and the Retreat from Kabul

One of the most spectacular examples of inept generalship was given in 1842 by Major-General William Elphinstone – better known to his men as 'Elphey Bey' – during the First Afghan War (1838–42). 'His pusillanimous conduct,' wrote Field Marshal Sir Gerald Templer, 'led to the most disgraceful and humiliating episode in our history of war against an Asian enemy up to that time.'

It had all begun three years earlier when the British, fearful for the security of India, had invaded Afghanistan and deposed the Amir, Dost Mohammed, on the grounds that he was moving towards an alliance with Russia. In his place they installed the malleable but unpopular former ruler, Shah Shuja. An uprising was inevitable. Lieutenant-General Sir John Keane, then British commander, recognized this as early as the autumn of 1839. 'I cannot but congratulate you for quitting the country,' he told a young subaltern, 'for, mark my words, it will not be long before there is here some signal catastrophe!'

At first, all seemed ominously quiet as soldiers sent for their families and the Kabul garrison settled down to the typical life of an Indian hill station, with concerts, horse races, skating and cricket matches. When Shuja's harem arrived from India, the British felt secure enough to vacate the citadel – the imposing Bala Hissar – to house them. It was the first of many blunders.

The removal of most of the troops to cantonments on the plain outside the city proved to be a tragic error. The ground selected was low and swampy, commanded on every side by hills and forts; between it and the city was a mass of orchards and gardens, crisscrossed with irrigation ditches and particularly unsuited for the speedy movement of troops and artillery.

The design of the cantonments, a rectangle 1,000 yards by 600, surrounded by a low rampart and a narrow ditch, could not have been worse. A defensive perimeter of almost two miles was far too big to be manned effectively by the garrison available, while a second compound half as big again jutted out from the northern rampart and so negated its purpose of defence. Even worse, the commissariat stores were placed in a fort 300 yards away.

The man ultimately responsible for the appalling site and design of the cantonments was Elphey Bey's predecessor, Major-General Sir Willoughby Cotton. Complacency is the only possible explanation, illustrated by Cotton's assurance to

Elphey Bey, on handing over command, that 'you will have nothing to do here, all is peace.'

Yet the warning signs were there. An uprising by the Durani clan was nipped in the bud in January 1841 thanks to the decisive action of Major-General William Nott, a tough, no-nonsense officer in command at Kandahar. But Nott was overlooked as Cotton's replacement in Kabul because of his abrasive nature and lack of respect for Shah Shuja. It did not help that he was an 'Indian' officer, commissioned into the army of the East India Company.

Elphey Bey, on the other hand, was a Queen's officer with seemingly impeccable credentials. A grandson of the tenth Baron Elphinstone, his father had been a director of the Honourable East India Company, his cousin Governor of Madras. He himself had served the Duke of Wellington with distinction in the Peninsular campaign and at Waterloo.

Yet he had been on half-pay for most of the intervening period, not returning to active service until 1839 when posted to India as commander of the Benares Division. Now almost sixty, flatulent and incontinent, with such bad gout that he could barely walk, he tried to refuse the onerous appointment. But Lord Auckland, the Governor-General of India, under pressure from the Horse Guards in London (the headquarters of the British Army), would not be moved. When Nott heard of the appointment he dismissed Elphey Bey as 'the most incompetent soldier that was to be found among the officers of the requisite rank'.

Elphey Bey finally arrived in Kabul in April 1841 and was immediately laid up with fever and rheumatic gout. By July he was showing little sign of improvement. 'I have it now in wrist, knee and ankle,' he wrote to his kinsman, Lord Elphinstone, 'and if ordered by the medical committee I shall apply to Lord Auckland to be relieved . . . My stay would be useless to the public service and distressing to myself.'

Events overtook him. In August 1841, Lord Melbourne's Whig administration was replaced by that of Sir Robert Peel's Tories. Fearing that the Afghan adventure would ruin the heavily indebted East India Company, Peel demanded economies. Sir William Macnaghten, the British Envoy in Kabul, responded in September by halving the subsidy paid to the eastern Ghilzais, the tribesmen who controlled the most direct route to India – through the Khyber Pass via Jalalabad. Their predictable response was to fall upon the next caravan from India. But Macnaghten, anxious not to delay taking up his new post as Governor of Bombay, played down the threat. The Ghilzais, he wrote to Lord Auckland, were simply 'kicking up a row about some deductions which have been made from their pay', and would be 'well trounced for their pains'.

Then, to save a few more pennies, he actually weakened the British presence in Kabul by insisting that Brevet Colonel Sir Robert Sale's brigade, at the end of its tour of duty in Afghanistan, march through the territory of the troublesome Ghilzais to meet its relief in Peshawar, rather than have the relief march in from India. By now, Auckland had at long last acceded to Elphey Bey's pleas that he was unfit to command and given him permission to return to India, while Nott would march up from Kandahar to replace him. But it was too late. En route to Peshawar, Sale's brigade was badly cut up by the Ghilzais, and before long the whole country was in open rebellion.

The explosion in Kabul came at dawn on 2 November with an attack on the British Residency. When word reached the cantonments, Elphey Bey proved incapable of decisive action. Twice he countermanded an order for his irascible second-in-command, Brigadier-General John Shelton, to march to the Bala Hissar on the grounds that Shah Shuja's troops had the disturbance in hand. In fact they had been ambushed in the narrow streets and by the time Shelton reached the citadel it was far too late to intervene. Sir Alexander Burnes, the Kabul army's Political Officer, is said to have been hacked to pieces while trying to escape in native dress.

A siege now began, and more than 4,500 British and East India Company troops, not to mention 12,000 camp followers, found themselves penned up in a wholly inadequate defensive position. Their predicament was made infinitely worse when the nearby commissariat fort fell to the rebels on 5 November.

All the while, Elphey Bey gave the impression of a drowning man. Ignorant of Afghanistan and its people, he sought advice from officers of all ranks and was 'in a constant state of oscillation; now inclining to one opinion, now to another; now determining upon a course of action, now abandoning it'.

On 6 November, just four days after Burnes's murder, he informed Macnaghten that the troops were low on ammunition and advised him to seek terms. The truth, according to Lady Sale, who was with the trapped British force, was that there was enough powder and ammunition 'for a twelve months' siege!'

At first Macnaghten ignored the General, preferring instead to try and bribe certain chiefs to forsake the rebellion. Meanwhile, desperate for reinforcements, Elphey Bey ordered General Shelton to march with his force from the Bala Hissar. It was a mistake: the army should have joined Shelton, not the other way around. And it did not help that Shelton made little attempt to hide his contempt for his vacillating superior, bringing his bedroll to the euphemistically named 'Councils of War' and often falling asleep.

Elphey Bey's despondency soon spread. 'The number of *croakers* in garrison became perfectly frightful,' wrote Lieutenant (later General Sir) Vincent Eyre, 'lugubrious looks and dismal prophecies being encountered everywhere.'

After a British attempt to clear Afghan artillery from

nearby heights ended in crushing defeat on 23 November, the rebel leaders offered a truce. Some of them were fearful that the destruction of the British army at Kabul would lead to another, more powerful, force being sent from India. Others, notably the Kuzilbashis, had no wish to see Dost Mohammed's line restored, and might well have sided with Shuja if the British leadership had displayed more backbone. Even Akbar Khan, Dost's son, was conscious of the need to tread warily in case the British took revenge on his exiled father.

Encouraged by Elphey Bey, Macnaghten agreed to the withdrawal of the British and the reinstatement of Dost in return for a face-saving guarantee that the Afghans would not ally themselves with another foreign power – by which he meant Russia. But Macnaghten then made the mistake of trying to double-cross the Afghans by renegotiating more favourable terms with those chiefs who feared a return to Dost's authoritarian rule. His duplicity was discovered by Akbar and the hardline chiefs who lured the Envoy out of the cantonments on 23 December, ostensibly to agree to new terms. Instead they murdered him and one of his political officers, and that evening his headless trunk could be seen hanging from a meat-hook in the bazaar.

Despite the fact that the abductions and murders took place within sight of the cantonments, no attempt was made to recue them. Elphey Bey later excused himself by saying that he had assumed the Envoy 'had proceeded to the city for the purpose of negotiating'. When the truth became known the following day, the dithering general was all too willing to accept Akbar's explanation that the unruly Ghazis were to blame. He and his senior officers had long since lost the stomach for a fight and seemed to feel that renewed negotiations were the only option.

It was the junior officers who thirsted for revenge. Major Eldred Pottinger, since Macnaghten's murder the senior 'political' (he took over as Resident), urged an immediate assault on Kabul which the troops, incensed by the murders, 'would no doubt have stormed and carried'. When Elphey Bey demurred in favour of negotiating terms, Pottinger argued that Akbar was not to be trusted and that the only way to save their honour and part of the army was either by marching to the Bala Hissar and holding out until spring, or abandoning their baggage and forcing their way through to Jalalabad. Unfortunately, a Council of War of senior officers agreed with Elphey Bey and decided that both courses were impractical.

On New Year's Day, as heavy snow fell, an agreement was signed with Akbar and the Afghan chiefs. The British would leave Afghanistan with an armed escort to protect them from the hostile tribes along the way. They would take with them only six artillery pieces and three mule-borne mountain guns; the rest, along with the muskets and ordnance stores in the magazine, would be left.

So on 6 January 1842, the once proud Army of the Indus marched forlornly out of the cantonments. Of a total of about four thousand troops, less than a quarter were Europeans, mostly members of the 44th Regiment and the Royal Horse Artillery. The rest were sepoys of the East India Company's army and members of Shuja's own infantry and cavalry. They were accompanied by 30 or so European women and children, and more than 12,000 camp followers. To reach their destination, they would have to march more than 80 miles through snow-covered passes held by hostile tribesmen.

Despite the assurances of protection, the struggling column was hounded from the start. Darting in like wolves, Afghan riders drove off baggage animals and slaughtered stragglers. Soon the snow was marked by a trail of bloody corpses and scattered belongings. By nightfall the march had covered just 6 miles. Much of the baggage had been lost and there were, Lady Sale noted, 'no tents, save two or three small palls. All scraped away the snow as best they could, to make a place to lie down on. The evening and night were intensely cold. No food for man or beast was procurable.'

By dawn, scores had died of exposure and many more were frostbitten. The Indian soldiers, in particular, were suffering because Elphey Bey had ignored Pottinger's suggestion that horse blankets should be cut up and used as puttees. A breakdown of discipline was inevitable. Long before the advance guard moved off without orders at 7.30am, hundreds of sepoys and camp followers had left and most of Shuja's troops had deserted.

That morning, the rearguard was heavily attacked and the three mountain guns abandoned in the confusion. Akbar and 600 horsemen then appeared, claiming to be the promised escort and blaming the column's misfortune on its premature start. Akbar also suggested that the column halt for the day to allow food and firewood to be brought up and the marauders to be dispersed. Incredibly, Elphey Bey agreed. 'Here was another day entirely lost,' commented Shelton, who bitterly opposed the halt, 'and the enemy collecting in numbers.'

With the temperature falling to ten below zero, the night was the worst yet. Past caring, many sepoys burned their equipment and even their clothes in an attempt to keep warm. The column had covered just 10 miles in two days.

Next morning, the demoralized procession reached the dreaded Khurd-Kabul Pass, 5 miles long and hemmed in by a sheer mountain face on either side. At first, all went well. The 44th Foot cleared the entrance to the pass at the point of the bayonet, while a cavalry charge dispersed a body of enemy horsemen hovering near by. But as the column – with its effective fighting strength now down to 1,000 men – moved into the pass, the waiting Ghilzais on the heights ignored Akbar's shouts and opened fire. All order disappeared as thousands raced forward in blind panic. Lady Sale was shot in the wrist but managed to gallop through to safety with her twenty-year-old daughter,

Alexandrina. Accompanying the irregular cavalry in the vanguard, they had donned turbans and poshteens, like the native troopers, to avoid drawing attention to themselves. Less fortunate was Alexandrina's husband, Captain Sturt, who was mortally wounded. In all, 3,000 bodies, mainly those of camp followers, were left in the pass.

Next day, when Akbar suggested taking under his protection the remaining families of the British officers, promising to bring them down in safety one day's march behind the army, Elphey Bey agreed. He also sent their husbands.

By the fifth day, with most of the sepoys suffering from severe frostbite, the Europeans were the only effective troops left. Two miles beyond their camp, in the narrow gorge of Tunghi Tariki, the Afghans attacked in force. The vanguard, made up of the 44th, some Company cavalry and the one remaining Horse Artillery gun, managed to break through, though with heavy casualties. The rest were slaughtered. Many had their throats cut as they lay defenceless in the snow. That night Shelton advised a forced march and Elphey Bey, for once, agreed. Now only 450 troops and 3,000 camp followers were left.

. . . the best marksmen in the world

The Afghan tribesmen who destroyed Elphey Bey's army in January 1842 were a mixed bunch. Mostly Pathans, true Afghans who claim descent from King Saul of Israel, they also included Hazaras, Tartar by origin, Parsiwans of Arab blood and Kuzilbashis from Persia.

Hardy hillmen, they made their living as soldiers, farmers or shepherds, leaving the 'lowly' work of trading goods to despised Hindus and other aliens. Religion

cont. overleaf

meant little to them and they were notoriously lax in their observance of the Muslim faith. When one young officer, commanding a troop of irregular Afghan horsemen, halted at the appropriate hour so that his men could turn to Mecca and pray, one trooper responded: 'You surely do not take us for clowns or pedlars; we are soldiers and never pray.'

Allegiance was owed only to feudal chieftains, who were loyally supported in the never-ending round of blood feuds and banditry. The scale of their fierce pride, love of independence and consequent hatred of 'foreign' interference was something the British were repeatedly to underestimate, with fateful consequences.

Murder was a fact of everyday life, and even family were not exempt. 'That is the grave of my father-in-law,' said one Afghan guide to a surprised British officer, pointing out a roadside memorial. 'I killed him shortly after my marriage, as his head was full of wind.'

Familiar with firearms from boyhood, the Afghans were natural fighters. They instinctively knew how to make the best use of cover and could move from rock to rock with the nimbleness of a mountain goat. Lieutenant Eyre, one of Elphey Bey's officers, described them as 'perhaps the best marksmen in the world'. They were also excellent horsemen – something of a necessity in a country that had no navigable rivers and was too rugged for wheeled traffic.

Their favoured weapon was a long-barrelled matchlock musket known as a *jezail*. Though of ancient design, and with a tendency for the barrel to burst, it had the crucial advantage of being able to shoot further than the shorter-barrelled muskets that were standard issue in the British and East India Company Armies. So frustrating did this handicap become at the Battle of Beymaroo Hills, where Elphey Bey's infantry squares were mown down with impunity, that the British officers were reduced to throwing rocks at their tormentors.

On 11 January, the remnants fought their way through to the village of Jugdulluk, and the following morning Elphey Bey accepted an invitation for him and Shelton to attend a conference in Akbar's camp near by. Negotiations dragged on all day as Akbar appeared to attempt to bribe a group of Ghilzai chiefs into giving the British safe passage. Assuring Elphey Bey that this had been arranged, he refused to allow the General to return to his dwindling command. In his absence, Brigadier-General Anquetil, now the senior officer with the column, decided to continue the march in the dark.

Finding the summit of the pass blocked with two barriers of prickly holly-oak, the men struggled desperately to find a way through. This only served to alert the Afghans, some of whom fired from the flanks while others rushed in with swords and knives. At last a hole was torn in the barrier, and men were shot by comrades and trampled by horses as a mad dash for safety ensued. 'The confusion was now terrible,' wrote Dr William Brydon, a surgeon formerly with Shuja's troops.

Just eighty men made it through the pass, including Brydon and fourteen other men on horseback. By dawn on 13 January, a week after setting out, the infantry reached Gandamak village. Between them they had a mere twenty muskets and two rounds apiece. Surrounded by hostile villagers, they agreed to parley, but were forced to fight when attempts were made to disarm them. Only six men survived the 44th Regiment's heroic last stand to be taken prisoner.

Meanwhile, Dr Brydon's party had ridden ahead and at Futtehabad, just 15 miles from Jalalabad and safety, they were foolish enough to accept the offer of food and rest. Betrayed by the villagers, just five escaped, and four of those were quickly overtaken and killed. Only Brydon rode on. Already badly wounded, he had to survive another three attacks, during which his bridle hand was cut and his horse shot in the groin, before he was spotted by a lookout in the British fort at Jalalabad. Of the 16,000 men, women and children who had set out from Kabul, only a handful reached safety; Brydon was the sole European.

The Afghan uprising was the result of a series of blunders by British political and military leaders. But once it had begun, only one man was responsible for turning a crisis into a catastrophe: Elphey Bey. An energetic and determined officer like General Nott would have saved the Kabul garrison, either by crushing the rebellion at its outset or by marching into the safety of the Bala Hissar. Elphey Bey attempted neither. Instead, his irresolution and timidity simply encouraged the rebels and lowered the morale of his own troops. Furthermore, when it was obvious that neither Akbar nor his followers could be trusted, he authorized a deal that led to the destruction of his army. He at least had the good sense to avoid a court-martial by dying in captivity.

Yet he deserves some plea in mitigation. He was neither physically nor mentally up to the job, and told Lord Auckland so. But instead of choosing a more suitable general like Nott, Auckland bowed to pressure from General Lord Fitzroy Somerset at the Horse Guards and appointed Elphinstone regardless. Later ennobled as Lord Raglan, Somerset was to play a central part in the events that led to the Charge of the Light Brigade and therefore, uniquely, had a hand in two of the greatest disasters in British military history.

Lord Raglan and the Charge of the Light Brigade

No military blunder better illustrates the amateurism of the early Victorian army than the Charge of the Light Brigade in October 1854; and no man was more guilty of that amateurism than Field Marshal Lord Raglan, the British commander in the Crimea.

In an age when commissions were purchased, and promotion was largely dependent upon wealth and influence, Raglan had the good fortune to be born the eleventh son of the fifth Duke of Beaufort. But he lacked the experience necessary to command an army. Most of his fifty years of service had involved staff work – first as aide-de-camp and then Military Secretary to the Duke of Wellington, then at the Horse Guards – and he had never commanded a formation larger than a battalion. His age, sixty-five, and the fact that he had lost an arm at Waterloo were hardly recommendations.

On the other hand, he could speak French, the language of diplomacy, and was noted for his equable temperament. Both attributes would help to maintain cordial relations with his French and Turkish allies (although his habit of referring to the enemy as the 'French', a legacy of the Napoleonic Wars, could not fail to ruffle a few feathers). It was also assumed that he had picked up a few tips in generalship from his great mentor, Wellington.

His senior officers had no such saving graces. Of six divisional generals, only two had commanded brigades in action, and only one, the 35-year-old Duke of Cambridge, was under sixty. He, however, was Queen Victoria's cousin and had never been to war. Even less deserving of their appointments were his five aides-de-camp: four were nephews and one a relation by marriage; none had any previous experience of staff work.

Then there was the problem of command. The 25,000-strong expeditionary force was made up of a jumble of semi-independent corps with only the infantry and cavalry under Raglan's direct authority. Furthermore, few of the individual regiments and battalions had experience of operating on even a brigade scale. The belief was still prevalent that battles should be fought in close formation, with repetitive drill taking precedence over battlefield tactics.

The Crimean War was sparked by a minor dispute over the guardianship of the Holy Places in Jerusalem, causing Russia to press her claim as protector of all Orthodox Christians within the Ottoman Empire. When the Turks refused to agree to this diminution of their sovereignty, hostilities began. But the deeper-lying cause of the conflict was Russia's belief in the imminent disintegration of the Ottoman Empire – famously described by Tsar Nicholas I as 'the sick man of Europe' – and her determination to be the first to pick over the bones (the most choice of which was the strategic city of Constantinople). Britain and France were equally determined to prevent Russia from expanding towards the Near East, for fear that it would upset the balance of power in the Balkans and allow Russian warships into the Mediterranean.

Initially, with the aim of intercepting the Russians before they reached Constantinople, the Allied troops were disembarked in Bulgaria. But when the Russians promptly withdrew, their respective governments decided on a more punitive course of action: to destroy the great Russian naval base of Sebastopol in the Crimea. The landings took place in September 1854 and within weeks the Allies had defeated the Russians at the Battle of the Alma and laid siege to the great port. However, the beaten Russian field army was soon heavily reinforced and the besiegers were in danger of becoming the besieged.

To supply their troops during operations, the British had

chosen the small harbour of Balaclava, 10 miles south-east of Sebastopol. It was situated on the open right flank of the Allied armies, and to protect it Raglan had posted the Cavalry Division in the plain to its north. He had also ordered the construction of a chain of redoubts along the Causeway Heights – a series of hillocks that ran east to west across the plain and divided it into two valleys: north and south. To the east of the plain was a steep escarpment known as the Sapouné Ridge, leading up to the Chersonese Plateau and, ultimately, to Sebastopol. It was from this plateau that the Allied armies were conducting the siege.

On 24 October 1854 – a week after the bombardment of Sebastopol had begun – Lord Raglan received word from a Tartar (native Crimean) spy that the Russians were preparing to attack Balaclava with 28,000 men. Yet he made no attempt to beef up the defence of his threatened supply port, despite the fact that two of the six redoubts on the Causeway Heights were unfinished and the rest were manned by just 1,400 Turkish militiamen who had never seen action, while only three of these redoubts were protected by artillery. Acknowledging the news with the words 'Very well', he simply requested that 'anything new was to be reported to him'.

This extraordinary oversight was partly the result of his general mistrust of spies, partly because there had been similar 'false alarms' the previous week, and partly because he was unwilling to denude the thinly spread force investing Sebastopol. Whatever the reason, it was to cost the cavalry dear.

The following day, an hour before dawn, the Russians attacked the Causeway Heights in overwhelming force. Within two hours all four manned redoubts had been taken, their inexperienced defenders fleeing in disorder across the south valley. Arriving at the edge of the Sapouné Ridge in time to witness this calamity, Raglan immediately issued orders for two infantry divisions to march down into the plain. Neither

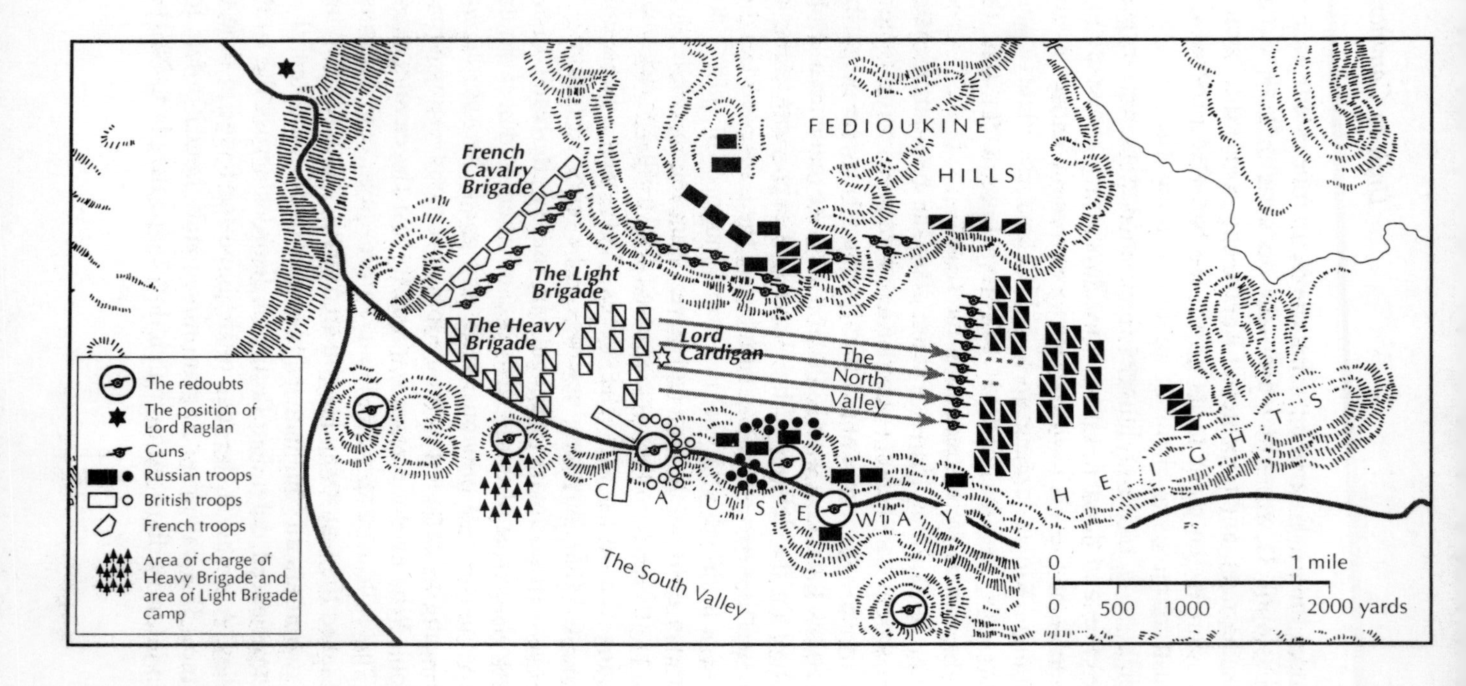
FEDIOUKINE
HILLS
French Cavalry Brigade
The Light Brigade
The Heavy Brigade
Lord Cardigan
The North Valley
CAUSEWAY
HEIGHTS
The South Valley
0 1 mile
0 500 1000 2000 yards
The redoubts
The position of Lord Raglan
Guns
Russian troops
British troops
French troops
Area of charge of Heavy Brigade and area of Light Brigade camp

would arrive in time to influence the course of the battle. Only the Cavalry Division of 1,500 sabres, a weak infantry battalion, the 93rd Highlanders, numbering 550 men, and a wavering battalion of Turks now stood between the Russians and Balaclava.

Watching helplessly from his elevated position, Raglan could see a huge force of Russian cavalry – more than 2,500 sabres in all – advancing up the north valley. It was obvious that their objective was Balaclava.

At this critical moment, William Howard Russell, the celebrated war correspondent of *The Times*, looked across at the amiable, one-armed Commander-in-Chief. 'Lord Raglan was by no means at ease,' he wrote. 'There was no trace of the divine calm attributed to him by his admirers as his characteristic in moments of trial. His anxious mien as he turned his glass from point to point, consulting with Generals Airey, Estcourt, and others of his staff, gave me a notion that he was in "trouble".'

Salvation came in the form of two immortal actions. First, the Highlanders – Russell's 'thin red streak tipped with a line of steel' – repulsed a charge of four squadrons of Russian hussars; then the Heavy Brigade charged the main body of Russian cavalry (three times its strength) and sent it reeling back over the Causeway Heights.

A curious lull now ensued as the Russians consolidated along three sides of the north valley. Their cavalry had withdrawn to the eastern end behind a battery of Cossack artillery, but their infantry still occupied the Causeway Heights and the Fedioukine Hills to the north. It was into this jaw that the Light Brigade would charge.

Raglan was anxious to take advantage of the defeat of the Russian cavalry by retaking the Causeway Heights, but there was still no sign of the infantry. He therefore sent the following order (known as 'the Third Order') to Major-

General the Earl of Lucan, commanding the Cavalry Division: 'Cavalry to advance and take advantage of any opportunity to recover the Heights. They will be supported by the infantry which have been ordered. Advance on two fronts.'

Having received the order, Lucan assumed that Raglan did not wish him to attack until the infantry had arrived – which was not the case. But it was a justifiable assumption given the relatively imprecise wording of the order and the fact that it was against all the rules of warfare to launch cavalry against infantry and artillery without support. Consequently, he ordered the Light Brigade into the north valley and positioned the Heavy Brigade in the south. They could then attack the Causeway Heights on two fronts when the infantry arrived.

As the minutes ticked by, Raglan became increasingly anxious. He was convinced that Russian morale had been dealt a heavy blow by the defeat of their cavalry and that an assault along the Heights would cause them to withdraw. But Lucan would not move. Suddenly, a watchful member of Raglan's staff exclaimed: 'By Jove, they're going to take away the guns!'

Sure enough, the Russians were bringing forward horses with lasso tackle to remove the naval 12-pounders that had been sited in three of the captured redoubts. Well aware that his hero, Wellington, had never lost a gun (to artillerymen, the capture of a gun was the equivalent of the capture of an infantry battalion's colours), Raglan turned to Major-General Sir James Airey, his Quartermaster-General, and dictated the following order for Lucan (known to historians as 'the Fourth Order'): 'Lord Raglan wishes the cavalry to advance rapidly to the front – follow the enemy and try to prevent the enemy carrying away the guns. Troop Horse Artillery may accompany. French cavalry is on your left. Immediate.'

Having checked the wording, Raglan handed the order to

Captain Louis Edward Nolan, Airey's aide-de-camp and the best horseman on the staff. It was an injudicious choice given Nolan's excitable nature and undisguised contempt for the cavalry commander. 'Tell Lord Lucan the cavalry is to attack immediately,' shouted Raglan as Nolan set off down the precipitous slope.

Within minutes, Nolan had located Lucan between his two brigades and handed over the order. Lucan was horrified. Now he was being asked to recover the guns without even infantry support. Seeing him hesitate, Nolan said sharply: 'Lord Raglan's orders are that the cavalry should attack immediately.'

On a much lower elevation than Raglan, Lucan could not see the guns being towed away. Could the Commander-in-Chief have a different objective? 'Attack, sir! Attack what? What guns, sir?' he demanded.

With an imperious gesture of his arm in the vague direction of the redoubts, Nolan retorted: 'There, my lord, is your enemy! There are your guns!'

But Lucan took Nolan's gesture to be towards the far end of the north valley where the Cossack guns were clearly visible, the sun glinting off their polished barrels. Assuming that Raglan intended him to charge down the valley, he ordered the cavalry to prepare for action.

Given that, two days later, he admitted that his task had been 'to prevent the enemy carrying away the guns lost by the Turkish troops', it is a mystery why he eventually sent the Light Brigade down the valley rather than along the Causeway Heights. One possible explanation is that a combination of Nolan's insolent gesture and the absence of any mention of the 'Heights' in the Fourth Order had led him to assume that Raglan intended him to advance down the valley as the only way to save the guns.

After all, Raglan's actual intention – an unsupported

cavalry charge against the enemy-held redoubts – was only marginally less suicidal. 'I do not understand,' wrote Russell, 'how the Light Cavalry could have succeeded in doing that which, it is said, Lord Raglan intended they should accomplish. The guns were in Redoubts Nos 1, 2 and 3. The first was plainly inaccessible to horsemen – to have charged 2 and 3 in the face of the force of Infantry, Artillery and Cavalry the enemy had within supporting distance of it, would have been quixotic in the extreme.'

With the die cast, Lucan made his final plans: the Light Brigade would lead the advance with the Heavy Brigade in support. But realizing that his despised brother-in-law, Major-General the Earl of Cardigan, the commander of the Light Brigade, was bound to object to such a mission, he decided to give the order in person. Riding over to where Cardigan was waiting at the head of his brigade, Lucan said: 'Lord Cardigan, you will attack the Russians in the valley.'

'Certainly, my lord,' came the reply, 'but allow me to point out to you that there is a battery in front, a battery on each flank, and the ground is covered with Russian riflemen.'

'I cannot help that,' Lucan retorted, 'it is Lord Raglan's positive order that the Light Brigade is to attack the enemy.'

Disdaining to continue the conversation, Cardigan rode over to Lord George Paget, commanding the 4th Light Dragoons. The brother of a man who had unsuccessfully sued Cardigan for sleeping with his wife, Paget had little time for his superior, nor the latter for him, but he was the next senior officer and his co-operation was essential.

'Lord George, we are ordered to make an attack to the front. You will take command of the second line, and I expect your best support; *mind, your best support*,' said Cardigan, pointedly.

'Of course, my lord,' replied an indignant Paget, 'you shall have my *best* support.'

Cardigan then returned to his position at the head of the brigade. Tall and handsome, in the striking blue and gold tunic, dark red trousers and fur busby of his old regiment, the 11th Hussars, he cut a magnificent figure astride his chestnut charger, Ronald. Behind him were his staff officers in their blue coats and cocked hats, themselves a little way ahead of the leading regiments, the 17th Lancers and 13th Light Dragoons, deployed side by side in two lines. Next came the 11th Hussars, 100 yards further back. Bringing up the rear, the same distance back still, were the 8th Hussars, on the right, and the 4th Light Dragoons.

'Here goes the last of the Brudenells,' Cardigan (whose family name was Brudenell) muttered as he turned to his trumpeter. 'Sound the advance!'

The 676 riders were still at the trot when Captain Nolan surged ahead of the first line, shouting and pointing his sword in the direction of the Causeway Heights. Realizing that the brigade was not going to wheel right to attack the redoubts, he was making a last desperate attempt to correct the blunder. Unaware of his intention, an infuriated Cardigan assumed he was daring to hurry the brigade along. 'No, no!' he shouted. 'Get back into line!' But Nolan was oblivious and continued his wild career across Cardigan's front, yelling as he went.

Then, with just 50 yards separating the two, a shell burst between them. Nolan gave a ghastly shriek, the sword dropped from his raised arm and his trunk contorted inwards in spasm. This convulsive twitch of his bridle hand caused his horse to turn and gallop back through the interval between the advancing squadrons of the 13th Light Dragoons. Seconds later his corpse slithered to the ground. A piece of shrapnel had hit him square in the chest, piercing his heart and killing him instantly. The last chance of averting the terrible tragedy had gone.

Onward the brigade rode into that terrible crossfire. 'Hell

had opened up upon us from front and either flank,' recalled Private Wightman of the 17th Lancers, 'and it kept open upon us during the minutes – they seemed like hours – which passed while we traversed the mile and a quarter at the end of which was the enemy. The broken and fast-thinning ranks raised rugged peals of wild fierce cheering that only swelled the louder as the shot and shell from the battery tore gaps through us, and the enfilading musketry fire from the Infantry in both flanks brought down horses and men.'

With just yards to go to the bank of white smoke that masked the Cossack battery, Cardigan raised his sword in the air and turned to shout a final command: 'Steady! Steady! Close in!'

As a last defiant salvo was fired, the front rank swept into and around the battery. While some gunners hid under their gun-carriages and limbers, others were ruthlessly sabred and speared as they tried desperately to hitch up and tow away their guns. Then the remnants of the brigade were rallied and led on in a desperate charge against the massed ranks of Russian cavalrymen beyond.

'It was the maddest thing that was ever done,' wrote a Russian officer. 'They broke through our lines, took our artillery, and then, instead of capturing our guns and making off with them, they went for us . . . They dashed in amongst us, shouting, cheering and cursing. I never saw anything like it. They seemed perfectly irresistible, and our fellows were quite demoralized.'

The Russian cavalry fled as far as a viaduct at the far end of the valley, a bottleneck that forced them to face their pursuers. Only now, realizing their vast superiority in numbers, did the hunted become the hunters. Cardigan, meanwhile, had retired to safety alone. Since then, however, 500 Russian lancers had ridden down from the high ground on both sides of the valley and formed up in front of the guns. Effectively cut off, the survivors of the Light Brigade still managed to hack their way out.

'We got by them without, I believe, the loss of a single man,' wrote Lord George Paget. 'How, I know not! Had that force been composed of English *ladies*, I don't think one of us could have escaped.'

When the battered remnants formed up near the same ground they had charged from just 25 minutes earlier, only 195 men were still mounted. Even when taking into account prisoners of war and those who returned alone, both men and horses, the losses were still staggering. Of a total of 294 casualties, 113 were killed and 134 wounded, with many others captured. In addition, 475 horses were killed or died of their wounds.

Speaking to a parade of the survivors, Cardigan said with some justification: 'This is a great blunder, but no fault of mine.'

So who was to blame? 'My opinion,' wrote Cardigan, 'is that [Lucan] ought to have had the moral courage to disobey the order till further instructions were issued.'

There is some truth in this, but all three principals bear some responsibility. By allowing his personal contempt for Lucan to get the better of him, Captain Nolan failed in the one essential duty of a staff galloper: to provide verbal clarification of a written message.

Lucan, on the other hand, failed to insist on a clear explanation of his objective. Instead, he seems to have come to the bizarre conclusion that Raglan intended him to save the guns by attacking the far end of the north valley rather than the Causeway Heights.

But Raglan himself must shoulder the bulk of the blame. Even interpreted accurately, his final order was both unnecessary and irresponsible. The guns had been spiked before capture and could not be fired, the infantry had nearly arrived, and an unsupported cavalry attack along the heights was bound to have been costly, if not disastrous. Yet having decided on this course of action, he should have taken into account the fact that Lucan's view of the battlefield was more limited than his and made his order more specific (by mentioning the 'Heights', for example).

Needless to say, Lucan and the dead Nolan were made scapegoats, with the former being relieved of his command after refusing to accept responsibility. Raglan, like Elphinstone before him, managed to avoid the repercussions of his mismanagement of the war in general – and the Battle of Balaclava in particular – by his timely death in the Crimea the following June. One officer described the cause as 'exhaustion from dysentery, added to, I hear, by worry of mind, poor fellow'.

McClellan at Antietam

The Battle of Antietam on 17 September 1862 – the bloodiest single day of the American Civil War – was a glorious opportunity for the North to strike a decisive blow and shorten the conflict. But owing to the vacillation of the Union commander, Major-General George Brinton McClellan, the chance was lost and General Robert E. Lee's Confederate Army slipped away to fight another day.

At the beginning of the month, fresh from his victory over Major-General John Pope's Union Army of Virginia at the Second Battle of Bull Run, General Lee had convinced Jefferson Davis, the President of the Confederacy, that the time was right to carry the war into the North. An invasion of the border state of Maryland seemed the best option. With many Marylanders already fighting for the South, there was the possibility that the presence of a Confederate army might persuade it to secede. More importantly, a victory in Union territory might encourage European recognition of the Confederacy and even force the North, reeling from recent defeats, to sue for peace.

On 4 September, to the strains of 'Maryland, My Maryland', Lee's exhausted but victorious Army of Northern Virginia, 45,000 strong, splashed across the Potomac River near Leesburg. Three days later, at Frederick, Lee issued a proclamation inviting Marylanders to flock to his standard. The response was disasppointing, however, and on 10 September he continued his advance north-east towards Hagerstown.

That same day, to secure his rear, he despatched Major-General Thomas 'Stonewall' Jackson with 25,000 troops to capture the large Union garrison and arsenal at Harper's Ferry, Virginia, on the confluence of the Potomac and Shenandoah Rivers. He was prepared to risk dividing his troops because he was convinced that Pope would need time to restructure his beaten army before pursuing.

But he reckoned without the organizational abilities of the new Union commander, the thirty-five-year-old Major-General George McClellan. On 5 September, Pope had been relieved and his Army of Virginia incorporated into the 90,000-strong Army of the Potomac. Although never formally given command, McClellan was put in charge of all Union forces in the vicinity of Washington and became the field commander by default.

Born in Philadelphia, the great-grandson of a general who had fought the British in the War of Independence, McClellan spent two years at the University of Pennsylvania before entering the US Military Academy at West Point in 1842. A brilliant student, the president of a society dedicated to the study of Napoleon's campaigns, he graduated second in the class of 1846 and was commissioned into the Engineers. Sent to serve in the war with Mexico of 1846–8, he promptly earned three brevet promotions for gallantry. In 1855, as part of a board of officers, he spent a year in Europe observing the different military systems – which included a visit to the Crimea to observe the siege of Sebastopol. Unfortunately, the lesson he learnt from that campaign was that a slow, methodical approach to warfare was necessary and rash attacks without proper support were to be avoided at all costs.

In 1857, he resigned his commission to become Chief Engineer of the Illinois Central Railroad and within three years was President of the Ohio and Mississippi Railroad. When the Civil War broke out in 1861, however, he was immediately appointed major-general in command of the Department of Ohio (including the states of Indiana and Illinois). His success in gaining possession of Western Virginia from the Confederates in the campaign of Rich Mountain resulted in his promotion to commander of the Potomac Division and ultimately Commander-in-Chief. But despite his administrative and strategic acumen, his tactical direction of the battlefield was too cautious. Consequently, his attempt to take Richmond during the Peninsular campaign of 1862 was a fiasco, despite vastly superior forces, and he was eventually recalled and his troops assigned to Pope.

Now after Pope's defeat at Second Bull Run, their roles had been reversed and McClellan immediately put his considerable organizational powers to good use. Within days of taking command, he had re-established the morale of the Union Army and on 8 September he set off from Washington in pursuit of Lee. His van entered Frederick on the 12th, in time to skirmish with Lee's rearguard, and the following day he benefited from an outrageous piece of good fortune. A private in the 27th Indiana Volunteers, which was camped on ground recently vacated by Brigadier-General D.H. Hill's Confederate Division, found a copy of Lee's entire plan of operations –

Special Order 191 – wrapped around three cigars. McClellan was ecstatic: 'Here is a paper with which if I cannot whip Bobbie Lee, I will be willing to go home.'

Aware that Lee had split his forces, McClellan abandoned his normal caution and ordered his army to force the passes over the South Mountain range in an attempt to drive a wedge between Jackson at Harper's Ferry and Lee near Hagerstown. But when the battle for the gaps began on the 14th, his normal prudence reasserted itself. Believing that the Confederate rearguard under General Hill had been reinforced by Major-General James Longstreet's corps, he drew back from ordering an all-out attack until his full strength had been deployed. In fact, Longstreet would not arrive until late in the day and a golden opportunity was thrown away.

That night, Lee withdrew 6 miles west to Sharpsburg, close to the Potomac ford, and the following day received the welcome news that Harper's Ferry had surrendered and Jackson was on his way to join him. Realizing the urgency, 'Stonewall' had set off with part of his command on a night march, with orders for two more divisions to set off on the 16th, and A.P. Hill's division to follow after sorting out the capture of 11,000 men, 13,000 small arms and 73 cannon.

Lee's plan to give battle near Sharpsburg along the Antietam Creek was a calculated risk. He was heavily outnumbered and if defeated his retreating army would be seriously impeded by the Potomac River. On the other hand, the South needed battlefield success in Maryland to discourage the North and impress European governments. It was for political reasons, therefore, that he could not return to Virginia without a tactical victory.

He was fortunate in that the opposing commander was McClellan. A more vigorous general – Lee himself – would have pursued the retreating Confederate army with all speed on the 15th, so as to be in position to offer battle the following day while Lee's forces were still divided. But McClellan wanted to be able to bring his full weight to bear and, other than some minor skirmishing and long-range artillery duelling, 16 September passed quietly. That afternoon Lee completed the concentration of his army (with only A.P. Hill's division still at Harper's Ferry). McClellan spent the day reconnoitring the enemy positions and formulating his battle plan.

By the morning of 17 September, Lee had about 38,000 men drawn up along a 3-mile front on rising ground to the west of Sharpsburg. They were well protected by the Potomac River to the rear and Antietam Creek to their front. But McClellan had 82,000 men at his disposal. It should have been enough.

His battle plan was simple but effective. He would attack in echelon, from right to left, and thereby force the numerically inferior Confederates to commit reserves to meet each fresh assault. When the right of the Confederate line had been sufficiently weakened, his left wing would launch the final and decisive attack.

The man chosen to open the battle was forty-year-old Major-General Joseph Hooker, commanding I Corps. A flamboyant and ambitious officer, nicknamed 'Fighting Joe', he was said to drink too much and have an insatiable appetite for women, particularly prostitutes. Reporters jokingly referred to the large number of the latter flocking to Washington in 1861 as 'Hooker's Division'. The name stuck.

Having crossed the creek the day before, Hooker's men advanced at dawn on 17 September towards the left of the Confederate line held by Jackson's corps in woods and fields to the north of Sharpsburg. Their objective was the German Baptist Dunker Church, on high ground to their front. The initial attack was successful and Jackson's two divisions were pushed out of the East Woods, through a large cornfield and into the West Woods behind. But with his line in danger of

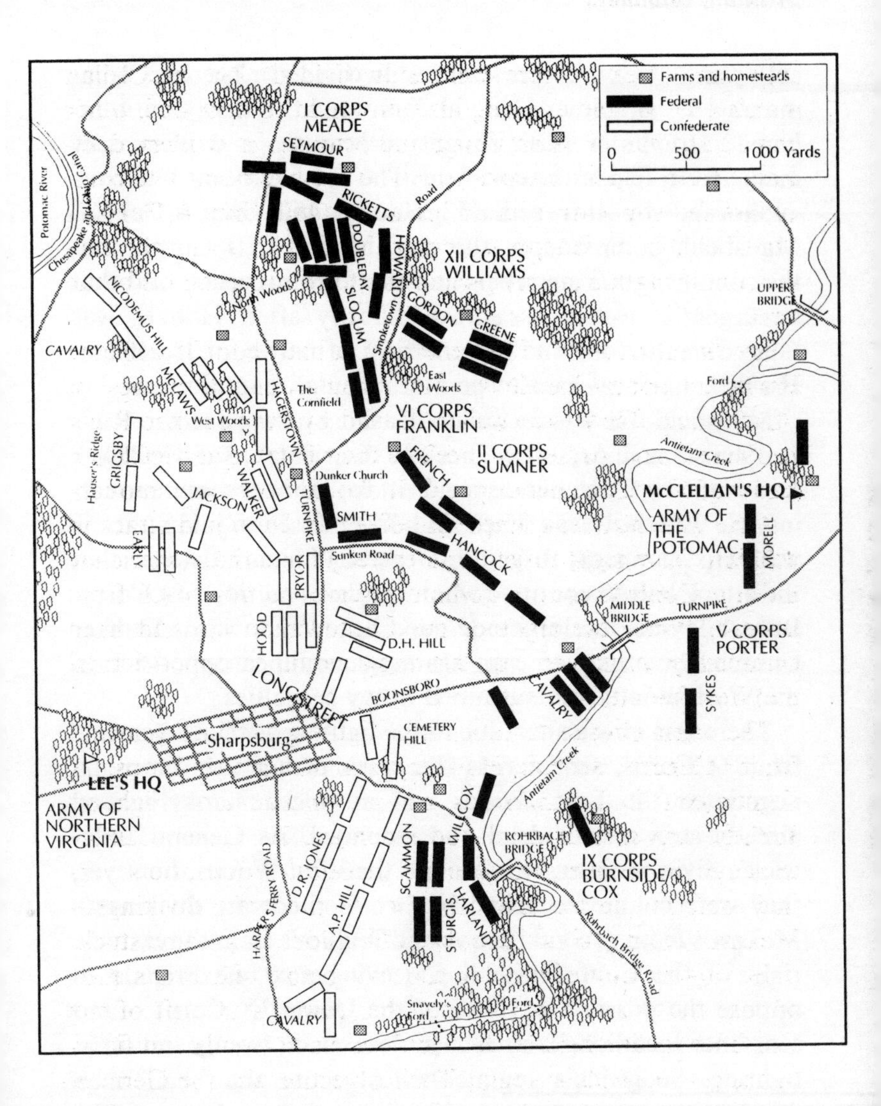

Farms and homesteads
Federal
Confederate
0
500
1000 Yards
I CORPS
MEADE
SEYMOUR
RICKETTS
DOUBLEDAY
HOWARD
XII CORPS
WILLIAMS
SLOCUM
GORDON
GREENE
Potomac River
Chesapeake and Ohio Canal
North Woods
Smoketown Road
Road
NICODEMUS HILL
CAVALRY
MCLAWS
HAGERSTOWN
The Cornfield
East Woods
West Woods
VI CORPS
FRANKLIN
II CORPS
SUMNER
FRENCH
GRIGSBY
Hauser's Ridge
JACKSON
WALKER
TURNPIKE
Dunker Church
SMITH
HANCOCK
EARLY
Sunken Road
PRYOR
HOOD
D.H. HILL
LONGSTREET
BOONSBORO
Sharpsburg
LEE'S HQ
ARMY OF NORTHERN VIRGINIA
CEMETERY HILL
D.R. JONES
A.P. HILL
HARPER'S FERRY ROAD
SCAMMON
STURGIS
WILLCOX
HARLAND
ROHRBACH BRIDGE
IX CORPS
BURNSIDE/
COX
Rohrbach Bridge Road
Antietam Creek
CAVALRY
MIDDLE BRIDGE
TURNPIKE
V CORPS
PORTER
SYKES
MORELL
ARMY OF THE POTOMAC
McCLELLAN'S HQ
UPPER BRIDGE
Ford
Snavely's Ford
Ford

collapsing, Jackson threw in his two reserve brigades commanded by Brigadier-General John Hood. Heavily outnumbered, Hood's Texans fought ferociously and eventually managed to regain the cornfield. The line had been stabilized.

Shocked by this setback, Hooker told Major-General Mansfield, commanding the neighbouring XII Corps, that the counter-attacking rebels had shattered his centre and that he urgently needed assistance. Shortly after, Hooker was injured in the foot and left the field of battle for treatment. His affliction was said to be more mental than physical.

Mansfield was less fortunate. Rushing forward to direct his two divisions in person, he became disorientated and mistook an advancing Confederate regiment for Hooker's men retreating. He was shot dead when his horse refused to jump a stone wall. His successor, Brigadier-General Williams, decided that the attack had to continue whatever the condition of I Corps. But while one division made good headway towards Dunker Church, the other came up against determined opposition in the West Woods and sustained heavy casualties.

There was now a short lull in the fighting as reinforcements from II Corps, sent across the creek earlier that morning, negotiated the East Woods and advanced across ground already strewn with dead and wounded. As General Sedgwick's division struggled through the West Woods, however, they were counter-attacked by two Confederate divisions – McLaw's from general reserve and Walker's from the extreme right of the Confederate line, leaving just one brigade to oppose the possible advance of the Union IX Corps – and sent into headlong retreat. In less than twenty minutes' fighting, Sedgwick's regiments lost more than half their effectives.

The Confederate advance, in turn, was stemmed by the timely arrival of two divisions from Major-General Franklin's VI Corps. Under the impression that the left of the rebel line was close to breaking point, Franklin was keen to attack. It may well have been a decisive manoeuvre because Lee had no reserves to hand. But Sumner, the commander of II Corps, urged caution and McClellan was of the same opinion. 'The commanding general came to the position,' wrote Franklin later, in remarkably restrained language, 'and decided that it would not be prudent to make the attack.'

Meanwhile, another of Sumner's divisions, led by Major-General French, had advanced towards a sunken road to the east of the West Woods defended by men of D.H. Hill's division (depleted by the fighting at South Mountain). Unaware of the strong defensive position until they were almost upon it, French's men suffered heavily from the initial exchanges. But they were reinforced, as were their opponents, and the bloody fight continued for more than three hours. Eventually, Union troops broke into the right of the Confederate position and began an enfilade fire down its length. The nearest rebel regiment panicked and fled, closely followed by the rest of the defenders. Only the fire from Longstreet's reserve artillery and a series of desperate counter-attacks, one led by D.H. Hill himself, prevented the Confederate centre from collapsing completely. It was just after one in the afternoon.

The man given the task of delivering the *coup de grâce* to Lee's tottering army was the thirty-seven-year-old Major-General Ambrose Burnside, the commander of IX Corps. Instantly recognizable by his luxuriant moustache and side whiskers, he is remembered today not for his military prowess but for the fact that his transposed name was the origin of 'sideburn'.

An apprentice tailor before entering the army as a lieutenant of artillery, he resigned his commission in 1853 to manufacture his own breech-loading rifle. Although the company foundered, his creditors continued to produce the rifle and more than 50,000 were issued during the course of the war.

Having rejoined the army at the outbreak of hostilities, he rose to prominence during the successful North Carolina expedition in early 1862. McClellan acknowledged his good work by giving him command of two corps – his own and Hooker's – during the early stages of the campaign. Yet McClellan chose to deploy these corps at opposite ends of the Union line at Antietam, thereby reducing Burnside's command to his original corps.

But no one told Burnside, and the confusion was made worse by the fact that General Reno, the temporary commander of IX Corps, had been killed at South Mountain. Consequently, no reconnaissance of the creek was carried out before the attack. If it had been, Burnside would have learnt of the existence of a lightly guarded ford a mile below the stone Rohrback Bridge that was to delay fatally his advance.

The intention was for Burnside's IX Corps to take the bridge over the creek during the morning, so that it would be in a position to advance against the weakened Confederate right – deployed in front of Sharpsburg – by noon at the latest. McClellan would then send in his uncommitted V Corps to complete the destruction of Lee's army.

According to the Union commander, the order for Burnside to 'carry the bridge, then to gain possession of the heights beyond, and to advance along the crest upon Sharpsburg', was despatched at 8am. But for some time nothing happened. An aide was sent to discover why, and reported back that 'little progress had been made'. Another was sent, and finally McClellan's Inspector-General, Colonel Sackett was despatched, with orders to remain with Burnside until his men were across the creek.

Burnside's version of events is that the first instruction was qualified with the words: '. . . await further orders before making the attack.' According to him, it was not until 10am that he received a definite order to advance. Contributing to the delay was his belief that Brigadier-General Cox was in charge of IX Corps, and therefore all orders had to go through him.

When the attack on the bridge did get under way, the defenders numbered just one brigade under Brigadier-General Toombs, a former senator with no military experience. But though vastly outnumbered, these Georgians fought tenaciously and, reinforced by a second brigade, managed to hold on for three precious hours. A further two hours were then wasted as Burnside insisted on sorting out his jumbled formations and resupplying them with ammunition. It was not until 3pm, a full two hours after the severe fighting on Lee's left and centre had concluded, that the final advance on Sharpsburg began.

At first, Burnside's men made good progress: the remnants of Brigadier-General Jones's division were brushed aside and it seemed as if nothing could prevent the Confederate line from collapsing. But in the nick of time, having just completed a 17-mile march from Harper's Ferry, A.P. Hill's division joined the fray. The fact that many of his men were wearing captured Union uniforms only added to the element of surprise. Taken in the flank, the leading regiments of IX Corps fled in disorder. Only stubborn defence by a brigade of Cox's Kanawha Division prevented the whole corps from being driven back over the creek. It was nearly sunset, and while some fighting continued after darkness, the battle was effectively over.

Three days later, without interference from McClellan, Lee completed the withdrawal of his battered but unbowed army back across the Potomac River into West Virginia. It could all have been so different had McClellan shown more vigour and less caution. The capture of Special Order 191 had given him the opportunity to engage and destroy Lee's formations one by one – yet he wasted it. He should have ordered a night march on 13 September and insisted on a vigorous pursuit after the battles in the mountains the following day – but he did neither. He should have attacked at Antietam on the 16th, before Jackson could arrive – but he was more concerned with studying the terrain and concentrating his own already superior forces.

When he eventually did attack, on the 17th, there was still the opportunity to defeat Lee. But McClellan's plan of battle called for careful timing and close supervision – and neither was exercised. The Commander-in-Chief remained on the eastern side of the Antietam Creek for most of the day, effectively allowing his corps commanders to fight their own separate engagements. The result was a series of uncoordinated assaults, with the attack of IX Corps taking place two hours after the fighting had ceased elsewhere. This delay – partly the result of poor communication between McClellan and Burnside – not to mention the timely arrival of A.P. Hill, cost the Union a great victory.

McClellan, however, was unrepentant. 'I feel that I have done all that can be asked,' he wrote to his wife on 20 September, 'in twice saving the country.'

But it was Abraham Lincoln's opinion that counted. On 13 October, irritated by McClellan's excuses for failing to pursue Lee across the Potomac, he asked: 'Are you not overcautious when you assume that you cannot do what the enemy is constantly doing?'

It was the final straw. On 7 November, McClellan was relieved of his command and never again employed in the field. His battlefield caution had cost the Union a victory which could have ended the Civil War in 1862.

General Warren and the Battle of Spion Kop

The Battle of Spion Kop on 24 January 1900 should never have been fought. That it was, and that it ended in unnecessary defeat for the numerically superior British, was largely thanks to Lieutenant-General Sir Charles Warren, arguably the most incompetent British commander of the whole Second Boer War of 1899–1902.

A month earlier, with the war just nine weeks old, the British commander in South Africa, General Sir Redvers Buller, VC, had made his first attempt to relieve the besieged town of Ladysmith in northern Natal by crossing the Tugela River at Colenso. It had ended in humiliating defeat (see p.85). In January, the British tried again, only this time the objective was Potgieter's Drift, about 20 miles upstream.

Catching the Boers off guard, the ferry was secured without a fight on 11 January and five days later Buller's vanguard had crossed and secured a bridgehead. But from his vantage point on Spearman's Hill below the drift, Buller could see the enemy busily entrenching Brakfontein Ridge, a section of the Tugela Heights, 3 miles north of the river. Convinced that by attacking such a strong position he would repeat the mistake of Colenso, he decided to force a second crossing 5 miles upstream at Trikhardt's Drift. There, his intelligence department assured him, the line of the river was held by just 600 Boers.

Buller therefore earmarked two-thirds of his army, under a separate command, to cross at Trikhardt's. Once established on the north bank, they would advance to the left of the Tabanyama Ridge, 3 miles to the north, and into the plain beyond. Once outflanked, the Boers would be forced to withdraw from their positions on the Brakfontein Ridge, leaving the road to Ladysmith open. A bold plan, it very much depended upon the commander at Trikhardt's Drift attacking with speed and resolution.

It was unfortunate, then, that the task was given to the fifty-nine-year-old Warren, described by one contemporary as 'dilatory yet fidgety, over-cautious yet irresolute and totally ignorant regarding the use of cavalry'.

Commissioned into the Royal Engineers, Warren began his long association with South Africa in 1876 when he was sent out to survey the boundary between the Orange Free State and Griqualand West, which had just been annexed by the Cape Colony. Ten years later, having helped apprehend the Egyptian killers of a British professor, he was appointed Commissioner of the Metropolitan Police in London. His controversial three-year tenure came to an end when his men failed to apprehend the notorious Jack the Ripper.

Rejoining the army, he commanded troops in Singapore and London before retiring in 1898. When war broke out the following year, however, he requested an immediate return to active service. At a subsequent meeting with Field Marshal Lord Wolseley, the Commander-in-Chief of the British Army, his wish was granted – though the two men violently disagreed about the right tactics to employ. In Wolseley's opinion, the only way to overcome the Boers in their entrenched positions was to outflank them; Warren, on the other hand, insisted that the British would have more success 'either by sweeping over them with very long lines of infantry' or 'by pounding away at them with artillery till they quailed'.

Despite this tiff, Warren was given command of the newly formed 5th Infantry Division and sent to South Africa as Buller's second-in-command (with a dormant commission to succeed him if he was killed or relieved). It says much for the loss of confidence in Buller that Major-General the Hon. Neville Lyttelton, one of the ablest brigade commanders in the Ladysmith relief force, begged Warren, shortly after he had arrived at Cape Town in December, to hurry north and 'back up Buller, otherwise the army would go smash'.

Buller, however, his confidence in shreds after a succession of disasters, was only too happy to give Warren the opportunity of commanding the next attempt to breakthrough to Ladysmith. If it succeeded, Buller would take the credit; if it failed, Warren could be blamed.

On 15 January, Warren was given his final instructions. After crossing Trikhardt's Drift with 15,000 men, part of his force would engage the Boers on the Tabanyama Ridge while the rest carried out the outflanking manoeuvre. Lyttelton, meanwhile, would make a diversionary attack against Brakfontein with his 4 Brigade.

Shortly after midnight on 17 January, Warren's vanguard came within sight of the drift. At that moment, barely 500 Boers were guarding the heights 4 miles away. Had Warren pressed on, his cavalry could have taken the heights that night – or certainly the following morning – and held them until infantry support arrived. But Warren was in no hurry, believing that a slow, methodical advance was the answer. His so-called 'flying column' stretched for 15 miles and took 13 hours to pass a given point. One exasperated officer wrote: 'We all wondered what was the cause of the delay. Some said folly, others incapacity, others even actual laziness.'

Although the water was shallow enough for cavalry to cross – and even infantry holding on to lifelines – Warren insisted on building two pontoon bridges (an operation he supervised in person) and only the cavalry and a single infantry brigade were on the northern bank by nightfall on the 17th. The Boers, meanwhile, were busily scraping together men to reinforce the threatened sector.

Only the British cavalry – 1,500 men under Colonel the Earl of Dundonald – showed any initiative that day. Skirting round the left of the Tabanyama Ridge, they brushed aside a small

Boer patrol and took up a strong position in the neighbouring hills. A short way ahead, up a gentle slope, was a pass that led to the Ladysmith plain. To outflank the entire Boer line, all Warren needed to do was follow up Dundonald's success by supporting him with infantry. Instead, fearing a trap, he recalled some of the cavalry. 'I had to make certain,' he later explained, 'that the mounted troops did not in the exuberance of their zeal get themselves into positions where they could not be extricated.'

The following day, ostensibly to protect the draught oxen grazing round the main camp, still more horsemen were pulled back. 'This order,' wrote Dundonald, 'paralysed the mounted brigade at the very moment that it needed strengthening.'

Incredibly, it was not until 19 January that Warren began his flank march. Part of the reason for the delay was that he would not move until the whole baggage train had been brought across the Tugela. Even when the column did finally move, it did not get far. Negotiating Venter's Spruit, a tributary of the Tugela, Warren felt vulnerable to attack and returned to the bridgehead. Buller, who visited him that day, was so shocked by his lack of progress that he even considered relieving him. 'On the 19th I ought to have assumed command myself,' he wrote later. 'I saw that things were not going well – indeed everyone saw that.'

That evening, at a conference with his senior officers, Warren made a fatal divergence from his original instructions. Pointing out that a flank march would dangerously lengthen his line of communication, he proposed taking the direct route over the Tabanyama Ridge – despite the fact that this meant making a frontal assault against prepared positions (now defended by 2,000 men) in direct defiance of Buller's orders. No one objected.

The following day the attack went in at dawn and made good early progress. Three Tree Hill and two other spurs that projected from the southern face of the ridge were occupied without loss. But as the attacking brigades attempted to cross the 1,000 yards of shallow, treeless gradient that led up to the summit, they were stopped in their tracks by a withering fire from the Boer trenches. By afternoon the advance had ground to a halt.

Fate now offered Warren yet another opportunity to make the all important breakthrough. Lord Dundonald, riding to the sound of the guns, could see Boers on a nearby ridge enfilading the British infantry attacking up the Tabanyama Ridge. Hoping to relieve the pressure, he sent a small detachment of his men up Bastion Hill, the leftmost spur running south from the ridge. They found the summit unoccupied, dug in and awaited reinforcements. None arrived because Warren was preoccupied with taking the eastern end of the ridge.

The attack resumed on the morning of the 21st and once again stalled in the face of accurate rifle fire. But by now the Boer defenders had been subjected to heavy artillery fire for three consecutive days and they were close to breaking point; during the afternoon a number of them began to slip away. 'All is lost if the British make an assault now,' read the diary entry of a German officer serving with the Boers. 'There is not a man in the trenches.'

Furthermore, the Boers were short of food and medical supplies, and many had gone down with dysentery from drinking infected water. Only the tireless example of their inspirational commander, Louis Botha, prevented the line from disintegrating. Nevertheless, Warren was on the verge of an undeserved victory when, incredibly, he called off the attack.

Next morning, Buller returned to Warren's headquarters on Three Tree Hill and berated his subordinate for his lack of success. 'I said that he must either attack or I should withdraw his force,' he wrote later.

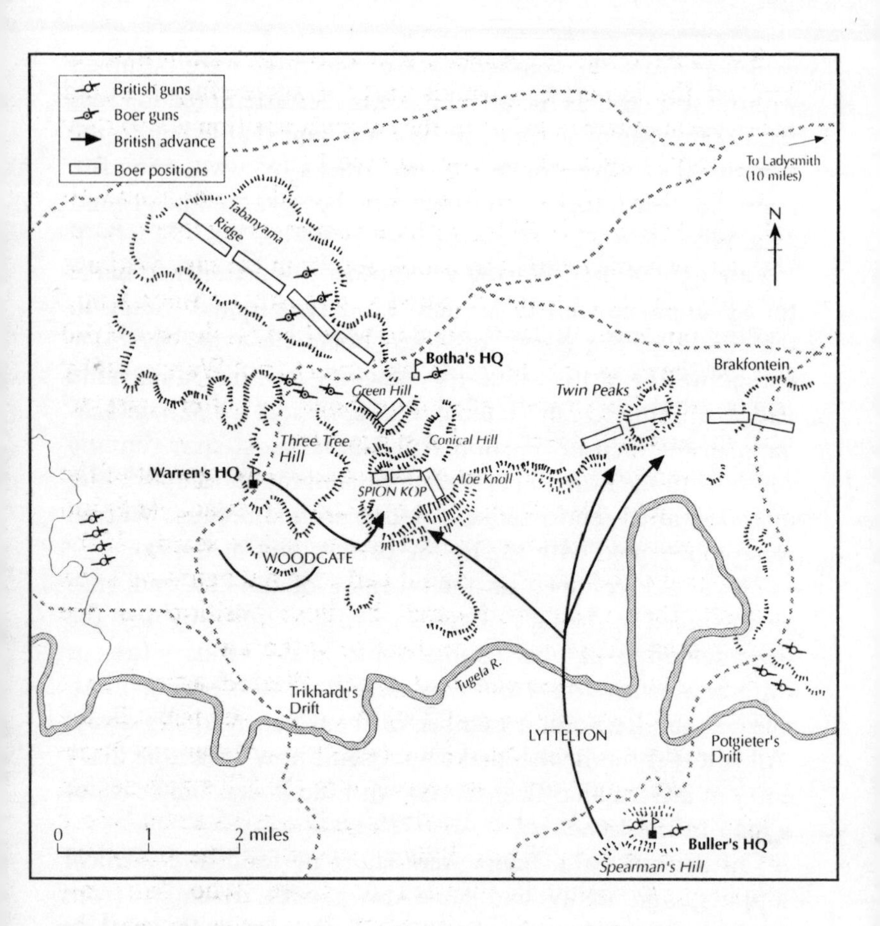
British guns
Boer guns
British advance
Boer positions
To Ladysmith
(10 miles)
N
Tabanyama
Ridge
Botha's HQ
Green Hill
Brakfontein
Twin Peaks
Three Tree
Hill
Conical Hill
Warren's HQ
Aloe Knoll
SPION KOP
WOODGATE
Tugela R.
Trikhardt's
Drift
LYTTELTON
Potgieter's
Drift
0
1
2 miles
Buller's HQ
Spearman's Hill

But Warren rejected Buller's advice to attack from Bastion Hill on the ludicrous grounds that, 'if successful, it would mean taking the whole line of the enemy's position which they [his troops] might not be able to hold.'

In any case, Warren continued, it would be impossible to take the Tabanyama Ridge without first securing Spion Kop, the steeply rising hill that overlooked it from the east. Without thinking, Buller replied: 'Of course you must take Spion Kop.'

With hindsight Buller claimed to have had his doubts. 'I did not like the proposal,' he wrote, and tried to put Warren off by 'saying that I always dreaded mountains, but after considerable discussion I agreed to his suggestion.'

However, the plan to capture the natural strongpoint of the whole range had its merits. Not expecting an attack up Spion Kop's precipitous slopes, the defence would be scanty. If the British could take the hill, hold it and drag field guns on to its summit, they would command the Boer positions on the Tabanyama Ridge and force them to withdraw.

But instead of moving that night, Warren agreed to a request from Major-General Talbot Coke – the man chosen to lead the assault column – for a twenty-four-hour postponement. Coke wanted time to reconnoitre the hill's approaches and, in any case, felt that his 10 Brigade needed a rest before going into action. When Buller learned of the delay the following morning, he insisted on Coke (who had only recently recovered from a fractured leg) being replaced by 'the more energetic General Woodgate' (who was merely lame). Major-General Edward Woodgate commanded 11 (Lancashire) Brigade.

Shortly after dark on 23 January, Woodgate's column of 1,700 men, which included a half-company of Royal Engineers, set off from a gully below Three Tree Hill, about six miles south-west of Spion Kop. Led by Lieutenant-Colonel Alec Thorneycroft and 200 members of his Mounted Infantry

– mainly *uitlander* refugees – the column reached the foot of Spion Kop at midnight. In pitch darkness and a light drizzle, they began the ascent up a zigzag goat path lined with boulders.

At exactly the same time, with the intention of shelling the bridges at Trikhardt's Drift, the Boers were struggling up the north slope with a 76mm Krupp field gun. A group of German engineers had been sent ahead to prepare a gun emplacement on the summit. They had joined seventy men of the Vryheid Commando in two shallow trenches at the top of the south-western spur, facing Trikhardt's Drift. Not expecting an attack, the Boers had failed to post pickets and most were asleep.

Shortly before 4am, Thorneycroft could feel the slope flattening out. It was getting lighter, but a dense mist had reduced visibility to a few yards. Hearing voices, Thorneycroft ordered his men to fix bayonets and form a line. As they neared the crest line, a lone voice called out: 'Wies daar?'

The leading files dropped flat as the Boers opened fire. But when the rattle of rifle bolts indicated that their magazines were empty, Thorneycroft leapt up and ordered his men to charge. Amidst cries of 'Majuba!', they poured into the enemy trenches, bayoneting one Boer while the rest fled, some without their boots. The British had taken the hill at a cost of just three wounded.

Arriving soon after, Woodgate ordered his sappers to construct a curved trench system, 300 yards long, on what seemed, in the mist, to be the forward crest of the summit. They did not bother to reconnoitre the gently sloping ground in front, and so were unaware that there was a second crest line a mere 200 yards ahead, and that 400 yards to the right the hill humped up into Aloe Knoll, a feature high enough for the enemy to pour enfilading fire from it into the entire position. Unknowingly, they had broken the cardinal military rule that there are only two places to hold on a table-topped hill: the forward crest, or a position immediately behind the rear crest. Instead, they were entrenched in centre of the plateau – the worst possible position.

As the half-company of sappers toiled away, equipped with just twenty picks and shovels, the infantry sat idly by. Unfortunately, the ground was solid rock and after three hours' digging they had only managed to construct a flimsy, 18-inch wall of stones and soil in front of a shallow ditch. Every member of the assault column should have been carrying an empty sandbag, to be filled with spoil from trenches and used to make a defensive parapet, but someone had forgotten to distribute them.

Botha, meanwhile, had been informed of the loss of the hill by panicked members of the Vryheid Commando. Determined to recapture it, he began to cast around for reinforcements and before long had dispatched 400 burghers – mainly from the Carolina and Pretoria Commandos – back up the hill. By 7.30 am, as the fog began to lift, the Boers were in possession of all the key points overlooking Woodgate's position: the forward crest, Aloe Knoll and Conical Hill, 800 yards to the north of Aloe Knoll, and only 100 feet lower.

Shortly after, the Boers opened fire on the exposed British positions, killing a number of men including a subaltern who was in the act of raising a sandwich of 'Gentleman's Relish' to his lips. Reacting quickly, the British officers led a series of bayonet attacks on the crest line and eventually succeeded in dislodging the defenders.

But as the sun burned the remaining mist away, the British realized the peril of their situation. The soldiers on the unprotected crest line, and those in the shallow trenches behind, were vulnerable to rifle fire from Aloe Knoll on the flank, Conical Hill to the front, and Green Hill to the front left. This wide arc of fire commanded every inch of the summit, and under the cover of that fire the Boers were able to retake the crest line after fierce hand-to-hand fighting.

They were immeasurably assisted by the supporting fire from three Krupp field guns, two 6-inch Creusots, and two pom-poms (quick-firing 1-pounders) that Botha had skilfully trained on the plateau. The British artillery, though ten times greater, was hampered by poor communications and the fact that it was unaware of the exact Boer positions. Only when its guns fired on Aloe Knoll did it have a material effect. Unfortunately, Warren was under the impression that his men occupied the whole plateau and ordered the fire to cease. As it was, many of the British shells crashed into their own positions and it may well have been 'friendly fire' that mortally wounded General Woodgate during the struggle for the crest line. Command devolved to Lieutenant-Colonel Blomfield of the Lancashire Fusiliers, but he was wounded soon after and Lieutenant-Colonel Malby Crofton of the Royal Lancaster Regiment took over.

His first action was to contact Warren on Three Tree Hill. As no telephone line had been run up the hill (though the equipment was available) and the remaining heliograph operator could not be found, Crofton discovered a Private Goodyear who could signal by semaphore, using flags. Unfortunately, Goodyear decided to improve upon Crofton's message – 'General Woodgate dead. Reinforcements urgently required' – and sent the following version: 'Reinforce at once or all lost. General dead.'

Warren received this seemingly desperate message at 10am. Having already sent off two reinforcement battalions, he now dispatched another, ordering General Coke to accompany it and assume command on the summit. His final instructions were: 'Mind, no surrender.'

He did not, however, make any personal attempt to gain a clearer picture of the battle, nor did he deploy any of his remaining 10,000 men to relieve the pressure on the summit. Instead, after organizing the transport of water, food and ammunition up the hill, he went for a nap.

Coke was suspected of having done likewise. Starting his ascent at 11am, he was hampered by his recently fractured leg and did not reach a ledge lined with mimosa bushes, 600 feet below the summit, until midday. He remained there for some time – unconcerned enough to fall asleep – only appearing below the southern edge of the plateau (but not on the summit proper) at 5.30pm. An hour later he returned to the ledge.

Meanwhile, the battle had continued to ebb and flow. At one point during the morning, the British re-established themselves on a stretch of the crest line. But after rallying the fleeing men of his Carolina Commando, Commandant Henrik Prinsloo led a determined counter-attack and retook it.

By midday, the focus of the fighting had switched to the main British trench. Surrounded by dead and dying men, its survivors were hungry, parched, low on ammunition and close to breaking point. At about 1pm, as hundreds of Boers crept closer, using boulders as cover, the will to resist disappeared and a row of handkerchiefs could be seen fluttering above the right section of the trench.

'The English were about to surrender,' recalled De Kock, a Transvaaler, 'and we were all coming up, when a great big, angry, red-faced soldier ran out of the trench on our right and shouted, "I'm the Commandant here; take your men back to hell, sir! I allow no surrenders."'

It was Thorneycroft, hobbling up (he had sprained his ankle) from further down the trench. Before the Boers could intervene, he and a number of men retreated to a line of rocks behind the trench and opened fire. Minutes later, joined by a company of reinforcements from the Middlesex Regiment, they charged and retook the right of the main trench, although the Boers managed to withdraw with more than 170 prisoners.

From his vantage point on Spearman's Hill, 4 miles to the south-east, across the Tugela, Buller recognized the gigantic figure of Thorneycroft retrieving the situation. He was, he decided, the obvious replacement for Crofton who, from the wording of his earlier message, seemed to have lost heart. Buller therefore signalled to Warren: 'Unless you put some really good hard fighting man in command on the top you will lose the hill. I suggest Thorneycroft.'

For once Warren agreed and he promptly telegraphed Crofton that Thorneycroft had been promoted to the local rank of brigadier-general in command on the summit. He failed, however, to inform General Coke.

But an even greater blunder was about to affect the course of the battle. At 10am, General Lyttelton – with his brigade still south of the Tugela, near Spearman's Hill – had received a message from Warren asking for assistance. On his own initiative, Lyttelton set off for Spion Kop. Noticing that other reinforcements were already climbing the hill, he detached a battalion of the King's Royal Rifles and sent it towards high ground to the east of Aloe Knoll known as the Twin Peaks. On hearing of this manoeuvre, Buller considered it too dangerous and countermanded it. But his order was ignored by the battalion commander and by 5pm, having sent the few Boer defenders packing, the Rifles were in possession of the peaks.

Nothing, it seemed, could prevent them from advancing along the crestline, taking Aloe Knoll and linking up with the troops on Spion Kop. Nothing but Buller. Despite the battalion's success, the Commander-in-Chief wanted it withdrawn, and that night he eventually got his way. Yet another opportunity to roll up the Boer flank had been thrown away.

Thorneycroft, meanwhile, was unaware that a second battle had flared up on the south-eastern slopes of Spion Kop as the two remaining reinforcement battalions tried to prevent the

Boers from pushing round his right flank below the southern lip of the plateau. At 2.30pm, he sent a runner with a message for Warren that ended: 'If you wish to really make a certainty of the hill for night you must send more infantry and attack enemy's guns.'

Intercepting the message on its way down, Coke added: 'We appear to be holding our own.'

Coke changed his mind later on, having visited the upper slopes where he found a 'scene of considerable confusion'. But he made no attempt to assess the situation on the summit proper, and was unaware of Thorneycroft's promotion. Instead, he returned to his former position and signalled Warren, suggesting a withdrawal. In response, Warren sent instructions for Coke to report the situation on the summit to him in person.

The truth of the matter was that as the sun went down, at 6pm, the battle was as good as won. Even the incompetence of the generals had not prevented the ordinary troops, at a cost of 500 casualties, from heroically frustrating every Boer attempt to sweep them off the summit. Twin Peaks was in their hands and threatening the key salient of Aloe Knoll. A mountain battery was soon to begin its ascent and huge numbers of reinforcements were available. Even the Boers were no longer prepared to dispute the outcome and after dark they slipped away one by one until their positions on the summit were deserted, although a number remained near the crest.

Thorneycroft, however, was unaware of this and shortly after dark he held a council of senior officers. He was shattered by the day's fighting and had heard nothing from Warren since his appointment to command. In the absence of any clear instructions, therefore, he suggested an immediate withdrawal. The officers were divided until Crofton cast the deciding vote in favour.

Soon after, Lieutenant Winston Churchill (recently escaped

from Boer captivity and commissioned into the South African Light Horse) arrived on the summit with a message for Thorneycroft from Warren: an attempt would be made to drag two naval guns up the hill that night; a working party was already on its way with thousands of sandbags.

'But the decision had already been taken,' Churchill recalled. 'He [Thorneycroft] had never received any messages from the General, had not had time to write any . . . The fight had been too hot, too close, too interlaced for him to attend to anything . . . So having heard nothing and expecting no guns, he had decided to retire. As he put it tersely: "Better six good battalions safely down the hill than a mop up in the morning."'

During the night, to their astonishment, a handful of Boers searching for the body of a comrade discovered that the hill had been abandoned. 'Instead of a great and terrible fight as we had expected,' wrote one, 'we found that through God's goodness and mercy, the enemy had taken such fear in the night that they had left their positions and had abandoned a big part of their dead and wounded.'

There were 650 British corpses – smelling of 'blood and brains' – in an area the size of a rugby field. More than 70 of them had been shot in the right side of the head, testimony to the destructiveness of the Boer position on Aloe Knoll. In addition, 554 men had been wounded and 170 captured. Given the high proportion of fatal casualties, one observer's description of the battlefield as 'that acre of massacre' is particularly apt. For their part, the Boers claimed to have lost just 58 dead and 140 wounded.

Rarely was a battle more disastrously mismanaged and unnecessarily lost. But despite the furore back home when the news of the disaster was reported, Warren was undismayed, telling a young officer 'that the Boers had had a severe knock at Spion Kop and were ready to run on seeing British bayonets'. He blamed Thorneycroft for the setback, even going so far as to suggest that he should be shot for abandoning his post!

Buller was only too happy to point the finger at Warren. In a secret dispatch to the War Office, he described Warren as a man 'who cannot command, as he can neither use his staff nor subordinates. I can never employ him again on an independent command.'

He was right not to. If Warren had been a more decisive and resourceful commander, the Battle of Spion Kop would never have been fought. But instead of following up Dundonald's successful advance on the 17th, when the road to Ladysmith was open, he insisted on concentrating his army and waiting for its baggage. Even after he had disobeyed Buller's instructions and ordered a frontal assault on the Tabanyama Ridge, he threw away another golden opportunity by refusing to reinforce Dundonald's men on Bastion Hill. The following day, with the Boers at breaking point, he called off the assault.

Finally, in desperation, he decided to attack Spion Kop. Despite a series of blunders, and the fact that only a third of his total force was ever engaged, the stoic fighting performance of his troops made victory certain. Yet it was thrown away because of his failure to retain overall control. Not once did he visit the battlefield in person, preferring instead to command at a distance of more than 2 miles with less than perfect communications. Crucially, he made no effort to communicate with Colonel Thorneycroft, the commander on the summit, until it was too late. Without clear direction from the top, Thorneycroft felt justified in making the fatal decision to withdraw.

On 6 March, just a few days after the relief of Ladysmith, Warren was posted to the relatively quiet backwater of Griqualand West. Four months later he returned to Britain and spent his remaining years dabbling in astronomy and running the 1st Ramsgate ('Sir Charles Warren's Own') Boy Scout Troop. If only he had always dabbled in such harmless pursuits.

General Stopford at Suvla Bay

The virtually unopposed landings at Suvla Bay in August 1915 presented the Allies with a fleeting opportunity to end the stalemate in the Gallipoli peninsula – where the Allies had landed in April – and knock Turkey out of the First World War. It came and went because of the lack of urgency shown by the inept Allied commander, Lieutenant-General the Hon. Sir Frederick Stopford, who spent much of the crucial early stage of the battle out of touch with his troops.

Ten months earlier, Turkey (or the Ottoman Empire as it was then called) had entered the war on the German side and immediately set about attacking Russia in the Caucasus. By late December, the hard-pressed Russians were calling for British action in the eastern Mediterranean which would distract the Turks. Although the request was out of date by the time it reached London – the Turks having been forced on the retreat in January – Field Marshal Lord Kitchener, the Secretary of State for War, was still eager to respond.

Worried that the Turks might turn their attention towards the Suez Canal, he jumped at the opportunity to defeat them and open up a southern supply route to Russia. Winston Churchill, the First Lord of the Admiralty, also espoused the opening of a second front and together they urged an amphibious attack through the Dardanelles, the narrow waterway separating the Gallipoli peninsula from Asiatic Turkey.

Convinced by Churchill that the Royal Navy, with French support, could force the Dardanelles 'by ships alone', and that no troops need be diverted from the Western Front, the British government gave its approval. On 19 February 1915, British and French warships began to bombard the outer forts guarding the waterway; marines landed unopposed and a single division might then have been able to take and hold the entire peninsula.

It took another month, however, before the navy was actually ready to enter the straits. Once again the bombardment was successful; ammunition for the Turkish shore batteries was nearly exhausted and there was virtually nothing to prevent the great ships from sailing through the Dardanelles the following day. Unfortunately, they ran into a line of mines on their return and two British battleships and one French were sunk. The navies would risk no more losses.

In the meantime, Kitchener had decided that an army could be spared to take Constantinople. Commanded by General Sir Ian Hamilton – who left London with no staff, no maps and no information about Turkish defences more recent than 1906 – it was made up of five divisions, only one Regular. The rest were untried colonial and Territorial divisions.

When Hamilton arrived at the entrance to the Dardanelles, however, he found that the transports had been wrongly packed, with the essential supplies at the bottom. So the entire expedition, having put the Turks on their guard, sailed back to Alexandria to repack and retrain. In the month it took to return, the Turks rushed reinforcements into the Gallipoli peninsula. When the landings finally took place on 25 April, the Turks had six divisions to the Allies' five.

Even so, by feinting towards the neck of the peninsula and actually landing on the southern tip, the Allies gained the element of surprise. For twenty-four hours there was an opportunity for them to break out of their separate beachheads, link up and advance up the peninsula. But bad luck (the Anzacs – Australian and New Zealand Army Corps – landed on the wrong beach) and a lack of drive and initiative among the generals ashore meant the chance was lost.

The pressure to achieve a breakthrough at Gallipoli was dramatically increased in May when the Russians were catastrophically defeated in Galicia by a combined Austro-Hungarian and German army and eventually forced to retreat more than 300 miles. Promised five of Kitchener's untried New Army divisions (the force of volunteers who responded to Kitchener's famous call to arms of August 1914), Hamilton opted to begin the offensive with the first three to arrive because he wanted to take advantage of light conditions on the night of 6/7 August – darkness as the troops approached the coast, followed by a rising moon allowing them to see once ashore – that would not recur until September. Two divisions – the 10th (Irish) and 11th (Northern), together forming the new IX Corps – were earmarked for Suvla; the other would reinforce the Australian and New Zealand forces in the existing beachhead at Anzac Cove on the east side of the peninsula.

The original plan, devised at Hamilton's GHQ on the Isle of Imbros, was for the battle to begin at Anzac Cove on 6 August with a diversionary attack on the right against Lone Pine, while the main effort was made on the left towards the Sari Bair Ridge. That night, the 11th Division would land on beaches to the south of Suvla Bay – 4 miles up the coast from Anzac Cove – and capture the surrounding heights before morning. These included Karakol Dagh and Kiretch Tepe Ridge to the north, and Tekke Tepe Ridge, Chocolate Hill, Green Hill and 'W' Hills to the east. After daybreak, the assaulting troops would be reinforced by the 10th Division and, if necessary, the whole corps would assist the Anzacs in the battle for Sari Bair Ridge. Speed, needless to say, was of the essence. Every hour wasted after landing would enable the Turks to bring in reinforcements.

The choice of corps commander at Suvla was, therefore, crucial. Kitchener suggested Lieutenant-General Bryan Mahon, who was bringing out his 10th Division, but Hamilton did not think he was 'up to running a corps'. Instead he asked for either Lieutenant-General the Hon. Sir Julian Byng or Lieutenant-General Sir Henry Rawlinson, both of whom were serving in senior commands in France. Kitchener could not agree. Neither could be spared, and in any case Mahon – one of the army's most senior lieutenant-generals – could hardly be expected to take orders from a junior officer. This ridiculous stipulation left just two possible candidates: Lieutenant-Generals Sir John Ewart and Sir Frederick Stopford. With Ewart felt to be physically unsuited to such an onerous task, Stopford was selected by default.

The younger son of the fourth earl of Courtown, Stopford had been educated at Eton and Sandhurst before joining the Grenadier Guards. Unfortunately, most of his distinguished military career had been spent in staff and administrative appointments – including the post of military secretary to Buller during the operations to relieve Ladysmith in the Boer War – and the largest formation he had commanded in battle was a composite half-battalion during the Ashanti expedition of 1895 (shades of Lord Raglan). Having retired from the active list in 1909, he was holding the honorary appointment of Lieutenant of the Tower of London when war broke out in 1914. Recalled, he was given responsibility for the training and organization of the First Home Defence Army. Now aged sixty-one, in poor health, with most of his experience theoretical, he could hardly have been a worse choice to command the critical Suvla operation.

Briefed on the plan at Hamilton's headquarters in late July, he responded: 'I am sure it will succeed, and I congratulate whoever has been responsible for framing it.'

But he soon changed his tune after discussions with Brigadier-General Reed, VC, his Chief of Staff, who had been influenced by his time on the Western Front and as a liaison officer with the Turkish Army during the Balkan Wars of 1912–13. Reed was convinced that a system of trenches could not be taken without heavy artillery support. As howitzers could not be landed during darkness, he did not see how Chocolate and 'W' Hills could be captured before dawn. Also, he did not believe the troops would be physically capable of marching as far as Tekke Tepe Ridge. Finally, as Chocolate Hill had to be attacked from the north, it would be better to reduce the troops' length of march by landing inside Suvla Bay.

All this made sense in theory, but in fact the Turks had just three battalions at Suvla, and their defences lacked barbed wire and machine-guns. Nevertheless, Hamilton's staff made the desired amendments, freeing Stopford from the original prerequisite of capturing all the high ground before dawn. Now the main objective was to secure Suvla as a base for a new northern zone; only if this priority was not threatened would the troops capture Chocolate and 'W' Hills and then support the Anzac push. Thus was one of the main purposes of the Suvla landing ignored: namely, to clear by dawn the guns on 'W' Hills that would hamper the Anzac operations. In addition, and against the advice of the Royal Navy, which feared uncharted shoals, one brigade would now land in Suvla Bay.

Stopford made one other fatal error. Instead of following Hamilton's advice and remaining at Imbros – from where he could keep in constant contact with the headquarters of the 11th Division until he went ashore, as planned, at dawn on 7 August – he insisted on being put aboard the sloop *Jonquil*, which lacked signalling capabilities. During the crucial early hours of the operation, therefore, he was both literally and figuratively 'in the dark'.

The landings duly took place in total darkness at around 10pm on 6 August. For the assaulting battalions of 32 and 33 Brigades, destined for 'B' Beach below the southern point of

Suvla Bay, everything went to plan. 'The landing was a complete surprise,' wrote Midshipman Eric Bush. 'There was no opposition and not a single casualty. A few rockets were fired, and one or two rifle shots rang out in the darkness, but that was all.'

While two battalions began to construct a defensive line to secure the right flank of the beachhead, another attacked Turkish positions at Lala Baba and Nibrunesi Point along the lower part of Suvla Bay (this was the first action seen by any unit of Kitchener's New Army). By midnight, at a cost of heavy casualties, it had overrun the Turkish posts and the remaining five battalions were converging on Lala Baba.

The troops, however, were showing ominous signs of fatigue; the need to preserve strict secrecy had meant that many were not told of their destination until *that evening*, with the result that they had spent the day in full training instead of resting. To make things worse, 34 Brigade's landing at 'A' Beach on the upper part of Suvla Bay was a fiasco. Not only had the Royal Navy's three destroyers anchored in the wrong place – 1,000 yards to the south of their intended stations, slap bang in the worst shoals – they had also anchored in the wrong order, leaving the battalions to land in the wrong positions relative to each other.

Three companies of the 11th Manchester Regiment were the first to reach the shore, although they landed well to the south of 'A' Beach. Undeterred, they left one company to guard their right flank and the rest headed north towards their objectives. By 3am, despite isolated opposition, they had advanced two miles along the Kiretch Tepe Ridge. Unfortunately, they were an exception. The craft carrying the 9th Lancashire Fusiliers and the remaining company of the Manchesters ran aground about fifty yards from the coast. After struggling ashore through deep water, they searched vainly for their objective, Hill 10, which lay directly in front of

'A' Beach. But they, too, had landed well to the south and were unable, in the dark, to reorientate themselves. It did not help that their officers and NCOs were being ruthlessly targeted by Turkish snipers.

Barely a mile to the south, 32 Brigade and half of 33 Brigade were concentrating at Lala Baba. Their task was to move in the direction of Hill 10, from where they would make the assault on Chocolate Hill. However, the senior officer was unwilling to add to the confusion ahead and ordered the troops to stay put. He should have marched on regardless and taken over the role of capturing Hill 10 from 34 Brigade; but Stopford's failure to emphasize the importance of taking the high ground that night had allowed his subordinate commander too much discretion.

Finally, at 3am, with 34 Brigade's two remaining battalions trapped by shoals and hours behind schedule, Brigadier-General Haggard, now in overall command at Lala Baba, sent four companies forward. They were caught up in a Turkish counter-attack to the south of 'A' Beach and never reached Hill 10. Meanwhile, the 34th's commander, Brigadier-General Sitwell, had come ashore; as dawn was breaking, he ordered the 9th Lancashire Fusiliers to take the hill. Misdirected towards the wrong feature, the attack was easily repulsed by enfilade fire from the real Hill 10, a quarter of a mile to the north. An hour later, having set up his divisional headquarters near Lala Baba, Major-General Hammersley ordered 32 Brigade to support the 34th. This time the advance was successful and, shortly after 6am, the 100 Turks holding Hill 10 were forced to withdraw.

By now, transports carrying the 10th Division's 31 Brigade and half of its 30 Brigade should have landed on 'A' Beach to assist in the capture of Kiretch Tepe. But the mishaps suffered by 34 Brigade had caused Stopford to switch the landing place to 'C' Beach, a short way to the north of 'B' Beach. When the navy then informed him that they had found a suitable landing place to the north of 'A' Beach, and therefore closer to the troops' objective, he refused to change his orders on the grounds that it would cause too much confusion and delay. However, the divisional commander, General Mahon, and his remaining brigade, arriving by ship at 7.30am, were sent to the navy's new beach on the north side of the bay and told to support the 11th Manchesters. The division remained divided for several days.

The first artillery units were landed at daybreak; around mid-morning, Hammersley issued orders for an immediate attack on Chocolate and Green Hills. They were not carried out because Brigadier Sitwell, the senior officer at Hill 10, was of the opinion that 32 and 34 Brigades were too tired, thirsty and demoralized to take part!

It did not help that Stopford, having sprained his knee the morning before, had decided not to come ashore, and therefore remained out of touch.

Incredibly, it was not until 5.30pm that the attack was finally launched by three battalions from the 10th Division and two from the 11th (33 Brigade). Although supported by artillery and naval fire, they did not manage to take Chocolate and Green Hills until after nightfall. Nevertheless, the largely intact battalions of 32 and 34 Brigades had nearly caught up and could now have continued the advance. But they were recalled and the night was spent consolidating the meagre gains. No attempt was made to capture the tactically crucial 'W' Hills to the east, where Major Willmer, the German local commander, had concentrated his troops.

A similar lack of initiative was handicapping operations to the north of the beachhead. The remaining battalions of the 10th Division under Mahon had reached the Manchesters' positions on Kiretch Tepe during the early afternoon. But they did not even try to push on to where the high ground joined the Tekke Tepe Ridge.

'There was no one in front of us but the enemy, of whose whereabouts or number we had no knowledge, and we must try to dig in, as the staff were of the opinion that we should be shelled in the morning,' wrote an officer of the 5th Royal Inniskilling Fusiliers. 'The soil was hard and rocky . . . We dug all night and when dawn broke we had little to show for our labours.'

So by nightfall on 7 August, the Turks were still in possession of all the tactically important high ground because there had been no serious British effort to take it. It now became a race to land enough troops to break the deadlock before the Turks could bring up their two reserve divisions from Bulair, at the neck of the peninsula. Ordered at 7am that day, they were thirty marching hours away and would therefore arrive late on the 8th or early on the 9th. There was, therefore, still a possibility of seizing the high ground and threatening the Turkish positions around Anzac Cove, but everything depended upon speed and determination.

Sadly it was everywhere absent. Consulting his brigadiers in the early morning of 8 August, Hammersley was told that the men needed more time to rest and reorganize! Not able to discuss the matter with Stopford, who was still aboard the *Jonquil*, he bowed to their wishes and issued orders to link up the various points on the outer perimeter. Mahon, on Kiretch Tepe, was no less hesitant and insisted that he needed artillery support in order to advance. After receiving these reports Stopford made no effort to see for himself; convinced that not enough artillery and supplies had been landed, he vetoed any attack until that evening at the earliest.

Hamilton, meanwhile, was becoming increasingly anxious at the lack of progress. At 11am, his headquarters reminded Stopford of the importance of gaining a foothold on Tekke Tepe Ridge. The corps commander's feeble response, half an hour later, was to issue a heavily qualified order: 'If you find the ground lightly held by the enemy push on. But in view of the want of adequate artillery support, I do not want you to attack an entrenched position held in strength.'

Yet Stopford already knew, from air reconnaissance reports, that the expected Turkish reinforcements had still not reached the plain behind Tekke Tepe. As soon as Hamilton saw a copy of Stopford's order, he realized that he had to intervene. But with his own destroyer out of action, he was forced to wait five hours for a replacement and did not arrive off the bay until after 6pm.

Meanwhile, one of his staff officers, Colonel Aspinall, sent from Imbros at daybreak to report directly from the beachhead, had reached Suvla at noon. He was accompanied by Colonel Maurice Hankey, a politician who had been sent out from London to witness the build-up. 'An entire absence of the expected bustle of a great disembarkation,' wrote Hankey. 'There seemed to be no realization of the overwhelming necessity for a rapid offensive, or the tremendous issues depending on the next few hours.'

The pair dashed off to confront Stopford on board *Jonquil*. 'Well, Aspinall,' said Stopford, 'the men have done splendidly.'

'But they haven't reached the hills, sir.'

'No, but they are ashore.'

Aspinall then pleaded for an immediate advance before the Turkish reserves arrived. Stopford refused, saying he had already issued the necessary orders. Argument was useless so Aspinall left for the naval flagship and sent the following signal to Hamilton: 'Just been ashore, where I found all quiet. No rifle fire, no artillery fire, and apparently no Turks. IX Corps resting. Feel confident that golden opportunities are being lost and look upon the situation as serious.'

Hamilton, of course, was already on his way.

At last, perhaps stung by Aspinall's urgency, Stopford went ashore, arriving at Hammersley's headquarters at 4pm. The troops, he said, 'must attack at once'. He then returned to the *Jonquil* and the welcome news that the Turkish reserves had still not arrived and that the 53rd (Welsh) Division was on its way. At 5.30pm, therefore, he gave definite orders for a simultaneous advance to capture the heights all round Suvla. But he left the timing to Hammersley who, in the interim, had scheduled an assault by 33 Brigade against 'W' Hills the following morning. Hammersley merely extended the assault to a general attack. A whole day had been wasted.

'Such Generalship defies definition'

In June 1915, two months before the landings at Suvla Bay, General Hamilton told Kitchener that the commander would need to be a man of stiff constitution because the fierce Gallipoli summer would tax the fittest of men.

Yet Kitchener's eventual choice, Lieutenant-General Stopford, was so feeble that he could hardly lift his dispatch case when he boarded the train for the coast. Aged sixty-one, in poor health, he had been brought out of retirement because of the war. The fact that he had never commanded troops in battle was similarly ignored.

On 22 July, Hamilton impressed upon Stopford that the high ground at Suvla 'should be captured by a major, forceful assault before daylight' on 7 August, and urged him to show bold and vigorous leadership.

Instead, with the landings under way and confusion threatening to undermine the whole operation, Stopford sat aboard his sloop *Jonquil*, delighted that his men had even got ashore. At the same time, Hamilton remained at Imbros, fretting for news. 'Both generals waited for victory or defeat as if the whole operation was a horse race,' wrote Major-General J.F.C. Fuller. 'Such generalship defies definition.'

It seemed impossible for Major Willmer's tiny defending force of 1,500 Turks to be able to hold back the invasion for the forty-eight hours it would take for reinforcements to arrive. Yet so unconcerned was Stopford with urging his troops forward that the Turks managed it with just half an hour to spare.

That same morning, 9 August, Hamilton went ashore to discover Stopford supervising the building of some shell-proof huts for himself and his staff. 'He was absorbed in the work,' wrote Hamilton, 'and he said that it would be well to make a thorough job of the dug-outs as we should probably be here for a very long time.' The last Allied forces left the Peninsula some five months later.

Hamilton reached the *Jonquil* at 6.30pm and told Stopford that the Tekke Tepe Ridge had to be taken that night as it would be the first point of contact with the Turkish reserves. Stopford replied that the 11th Division needed another night's rest before attacking. Beside himself with rage, Hamilton went to see Hammersley, who reeled out the usual excuses: the troops were scattered, the terrain was difficult, no reconnaissance had been made. Finally, he agreed to order 32 Brigade, thought to be concentrated in the Suljik area, to attack that night so that at least one battalion would be on the ridge by dawn.

Unfortunately, two battalions of the brigade were holding key positions on the perimeter, and to rendezvous at Suljik at 10.30pm they had to march some distance. To make matters worse, the brigade commander would not move until all the battalions were concentrated. One never arrived, and after five and a half hours of pointless delay the advance finally began. The vanguard, after heavy fighting, eventually reached the top of the ridge, but was driven back by the sudden arrival of one of the Turkish reserve divisions.

A general attack, supported by the newly arrived 53rd Division, was put in later that morning and continued throughout the day. It achieved almost nothing now that the Turks, well entrenched on the high ground, had been strongly reinforced. Stopford, having finally landed his headquarters, spent this crucial day not directing the battle but supervising the building of some splinter-proof huts for himself and his staff.

The moment was past; there would be no break-out. Two costly beachheads had merely become three. The lack of resolve shown by Stopford and his senior commanders on the 8th had proved fatal. Eight days later, Stopford was relieved of his command and returned home to his post as Lieutenant of the Tower. Exonerated by a committee of general officers set up to investigate his conduct of operations, he never again commanded troops in the field and retired in 1920.

The Gallipoli campaign was finally abandoned at the end of the year on the orders of Sir William Robertson, the new Chief of the Imperial General Staff, who believed that the war could only be won in France and that Gallipoli was a distracting and costly sideshow (final casualties were in excess of 200,000). Suvla Bay and Anzac Cove were evacuated in late December, Helles Bay on 8 January, 1916. The withdrawal was a great success and not a man was lost, although vast quantities of stores had to be left. 'The Gallipoli expedition,' wrote A.J.P. Taylor, 'was a terrible example of an ingenious strategical idea carried through after inadequate preparation and with inadequate drive.' No general showed less drive than Stopford.

General Percival and the Fall of Singapore

The loss of the island fortress of Singapore to the Japanese in February 1942 was a disaster of quite monumental strategic and economic consequences. At a stroke, Britain forfeited its strongest foothold in the Far East, its major port for the export of much-needed rubber, as well as valuable naval and engineering facilities, military equipment, stores and fuel, two new capitalships and thousands of military and civilian lives. Such a catastrophe could only be the result of a catalogue of political and military blunders. Nevertheless, one man stands supremely responsible: the British commander in Malaya and Singapore, Lieutenant-General A.E. Percival.

As early as 1925, the three armed services had disagreed over the best way to protect such an important base. While the Army and Royal Navy wanted fortifications and heavy guns with which to repel an assault from the sea, the Royal Air Force argued for more aircraft in order to deter an attack before it came within range of the island. Predictably, the senior services got their way; the unfortunate consequence was that Singapore was left undefended on its northern – landward – side, the point at which it was linked to the mainland of Malaya by a causeway.

This omission began to look increasingly ominous in the summer of 1941 – almost two years after the outbreak of the Second World War – with Winston Churchill's decision to pour all available reinforcements into the struggle for North Africa, despite his military advisers' reminders that the defence of Singapore should be the main priority after the defence of Britain itself. This neglect was not helped when the US President, Franklin D. Roosevelt, and Churchill chose to punish Japanese aggression in French Indo-China in July by imposing a trade embargo. Cut off from the raw materials she needed to prosecute her war with China, Japan's only recourse was to extend her conquest of South-East Asia to Malaya (rubber and tin) and the Dutch East Indies (oil). As Britain and America would never allow this, war with these two powers was inevitable.

It duly began on 7 December when Japan launched her pre-emptive strike against the US Pacific Fleet at Pearl Harbor in Hawaii. At about the same time – early morning, 8 December, local time – advance elements of General Tomoyuki Yamashita's 70,000-strong Twenty-Fifth Army had landed at three points on the eastern neck of the Malay peninsula. Two were in Thailand, the third (and smallest) just inside the Malayan frontier at Kota Bharu.

The man charged with the defence of Malaya and Singapore was the fifty-three-year-old Lieutenant-General Arthur Percival, a former Assistant Chief of the Imperial General Staff. A businessman before the First World War, he had joined the Essex Regiment in 1914 and served with distinction in the trenches, winning the Distinguished Service Order, the Military Cross and the French Croix de Guerre in the process. He had remained in the Army after the war and by 1939 had risen to the rank of brigadier. More promotions followed and in May 1941 he was appointed GOC Malaya.

Of the 88,000 men that made up his new command, many were under-equipped and poorly trained. Yet Percival made little effort to rectify these shortcomings because he did not anticipate meeting the Japanese in the field. Like other so-called 'experts', he believed that the Japanese Army would be unable to advance through the Malay peninsula's supposedly impenetrable jungle, and certainly not with tanks. This might explain why, shortly before the invasion, thousands of War Office pamphlets on how to combat enemy tanks were found undistributed at Percival's headquarters.

'One can only assume that the anti-tank leaflets were left to moulder in a cupboard,' wrote Norman Dixon in his *On the Psychology of Military Incompetence*, 'because they were tactless enough to proclaim a heresy.'

On the day of the invasion, Percival issued the first of what seems to be a series of deliberately misleading war communiqués from his headquarters in Singapore, stating that the Japanese had failed in their attempt to land at Kota Bharu. A second, issued soon after, read: 'All [Japanese] surface craft are retiring at high speed, and the few troops left on the beach are being heavily machine-gunned.'

In fact, the 5,500 Japanese troops had little difficulty getting ashore and within hours were in control of the nearby airfield. The only boats 'retiring at high speed' were those going to pick up more troops. By 10 December, the Japanese 5th Division, having landed further up, had crossed to the west coast of the peninsula and had penetrated into Malaya on two fronts.

That same day, attempting to intercept non-existent Japanese troop transports off the east coast of the peninsula, the battleship *Prince of Wales* and the battle-cruiser *Repulse* were sunk by Japanese bombers. The officer in command, Admiral Sir Tom Phillips, who went down with his ship, had been warned that the RAF could not provide air cover but he sailed on regardless, comforted by the long-held Admiralty view that capital ships could not be sunk by aircraft. With the loss of these ships, and with only a small and ill-equipped RAF presence, the British were powerless to prevent the Japanese from landing more troops as and where they pleased.

From that date, the Commonwealth and Imperial forces were in almost constant retreat. But while men and vehicles stuck to increasingly clogged roads, they were constantly surprised by the appearance of Japanese tanks and artillery that had manoeuvred through well-spaced rows of rubber trees, and infantry that had marched through the bordering jungle. When the commander in northern Malaya, General

Heath, tried to make a stand on the Perak River, his line was turned by a Japanese column advancing obliquely across the peninsula – exploding the myth that the spine of mountains running north to south would prevent such a manoeuvre. The Japanese avoided another strong position at Kampar simply by moving troops further down the west coast in small boats captured during the advance.

By the beginning of January the British had been pushed back to a line just north of the Slim River, covering the Selangor province and the airfields of the Malayan capital, Kuala Lumpur. But the position was penetrated on the night of the 7th by a single company of tanks, which then raced on to seize a road bridge 20 miles behind the front line. Four thousand troops north of the river were captured; the Japanese lost six tanks and a handful of infantry.

That day, General Sir Archibald Wavell arrived in Singapore en route to Java to take up the new emergency post of Supreme Commander, Allied Forces, South-West Pacific. He decided to accelerate Percival's gradual withdrawal and base the new defence line on the southern city of Johore. Kuala Lumpur was abandoned on the 11th, Tampin two days later. But this simply enabled the Japanese to take advantage of the better road system in the southern province by deploying two divisions simultaneously, instead of in turn. The retreat was far quicker and more disorganized than had been intended, and it was impossible to construct an effective defensive line.

It was a similar story on the east coast, with Kuantan and its airfield evacuated on 6 January, Endau on the 21st. By the 30th, all British, Commonwealth and Imperial forces had been driven back into the southern tip of the peninsula. The rearguard crossed the 1-mile strait on to Singapore Island the following night. It had taken just fifty-four days for the Japanese to conquer Malaya. To their 4,600 casualties the British had suffered 25,000 (mostly prisoners).

'They're facing the wrong way!'

With its vast resources of rubber and tin, Malaya was one of the jewels in the British Empire's crown after the First World War. To protect it, the British government decided to construct a chain of airfields down the peninsula and a naval base below it on the island of Singapore, which would also serve to defend the vital sea supply routes between Britain and the Far East. But with money scarce and Japan, the main threat, seemingly passive, little was done in the 1920s.

All this changed in the early 1930s when Japan overran Chinese Manchuria, resigned from the League of Nations and denounced the restrictions of the international naval disarmament treaties. For Britain the construction of the Singapore naval base now became a priority. But because the military authorities did not believe a Japanese attack down the Malayan peninsula was possible, all the fixed 9.2-inch and 15-inch guns equipping the coastal batteries were pointing out to sea. The north of the island opposite the peninsula was left undefended.

The theory was that Singapore would be able to hold out for seventy days, the time it would take for a fleet to arrive from Britain. By the late 1930s, however, the Army had begun to concede that amphibious landings on the east coast of Malaya were possible. A paltry sum of money was therefore made available to construct a line of defences in southern Malaya that would prevent Japanese artillery from coming within the range at which it could fire on Singapore island. It had only been partly completed by the time of the Japanese invasion in December 1941. With the guns pointing the wrong way, the island was virtually defenceless when the Japanese attacked across the Johore Straits two months later.

During the ever-quickening British retreat down the peninsula, the Commander, Royal Engineers, Brigadier Ivan Simson, had repeatedly urged Percival to erect defences on the unprotected northern side of the island. Each plea was ignored. With the Japanese advancing ever closer, Simson decided to make one last attempt. He told Percival that he had the staff and materials 'to throw up fixed and semi-permanent defences, anti-tank defences, underwater obstacles, fire traps, mines, anchored but floating barbed wire, methods of illuminating the water at night'. Still Percival refused to give his permission.

'Sir,' continued Simson, 'I must emphasize the urgency of doing everything to help our troops. They're often only partially trained, they're tired and dispirited. They've been retreating for hundreds of miles. And please remember, sir, the Japanese are better trained, better equipped, and they're inspired by an unbroken run of victories . . . and it has to be done now, sir . . . once the area comes under fire, civilian labour will vanish.'

Percival was unmoved. 'Look here, General,' said Simson, no longer bothering to control his anger, 'I've raised this question time after time. You've always refused. What's more, you've always refused to give me any *reasons*. At least tell me one thing – why on earth *are* you taking this stand?'

After a pause, Percival replied: 'I believe that defences of the sort you want to throw up are bad for the morale of troops and civilians.'

Simson's blood froze. He could not believe what he was hearing. Singapore was as good as lost. 'Sir,' he said in a weary voice, 'it's going to be much worse for morale if the Japanese start running all over the island.'

Shortly after, Percival was visited by General Wavell who was 'very much shaken that nothing had been done' and, 'speaking with some asperity', demanded to know the reason. The answer was the same: to preserve civilian morale. It would be very much worse for morale, Wavell replied, if the troops in the peninsula were driven back on to the island.

Informed by Wavell of Percival's failure to protect Singapore, Churchill issued a directive on how to defend the north shore. Many of the measures were those advocated by Simson. Incredibly, Percival still refused to act. When he did finally issue a plan it was too late; the necessary civilian labour could no longer be found.

Norman Dixon offers a pyschological explanation for this extraordinary behaviour: 'In the case of Percival and Gordon Bennet [commander of the 8th Australian Division], to erect defences would have been to admit to themselves the danger in which they stood. In other word, their professed anxiety about civilian morale was really displaced from anxiety about their own morale. Looking further into the story of Singapore one is struck by the compulsive element in this refusal of the military to defend itself. Such compulsive behaviour is typical of many who present an authoritarian personality and are "reared" in an organization which traditionally deals with fear and danger by ritualistic means.'

But Percival was not content with failing to construct adequate defences. With his army of 85,000 troops now concentrated on the island, he decided to defend the coastline, spreading his forces weakly along an extended front rather than retaining a strong central reserve that could be rushed to any threatened point. He also shifted the majority of the defence stores to the north-east of the island despite strong indications that the attack would come from the north-west. When a build-up of Japanese troops made this certain, he ordered the stores to be moved back. But it was too late.

During the night of 8 February, over a front of 8 miles, two

Japanese divisions crossed the mile-wide channel between the mainland and the island in a motley collection of landing craft. Some were sunk, but most arrived safely, aided by yet more British blunders. The beach searchlights were not used, communications failed and the artillery was slow to put down its curtain of defensive fire. At daybreak, 13,000 Japanese were ashore and the three Australian battalions guarding the sector had fallen back inland. By midday, more than 20,000 had landed in the north-west sector of the island.

Numerically, the British had more than enough troops to repel the invaders, for no more than 35,000 Japanese soldiers were ever deployed on the island at any one time, although fresh troops arrived as replacements. But thanks to Percival's incompetent generalship, his men now had the handicap of poor morale to add to their lack of training. Lack of protection from the enemy's incessant air attacks lowered morale further, as did the burning of oil tanks to the defenders' rear. Churchill's appeal for the battle to be 'fought to the bitter end', and for commanders to 'die with their troops' for 'the honour of the British Empire', was the final straw. Most came to the reasonable conclusion that captivity was preferable to death. If only they had known what lay ahead.

The end came during the evening of 15 February. With the Japanese having entered the suburbs of Singapore city, food stocks low and the water supply about to be cut, General Percival went out under a white flag and surrendered in person. During the campaign as a whole, the British suffered more than 138,000 military and civilian casualties (including Indian, Australian, Malayan and Malayan-Chinese troops) – the vast majority of them taken prisoner.

The loss of Singapore was soon followed by the Japanese conquest of much of the rest of South-East Asia, including Burma and the Dutch East Indies. But the longer-term consequence of its fall was the destruction of the myth of white superiority. General Percival, therefore, by his disastrously incompetent and dogmatic handling of the campaign, played a small but significant part in encouraging post-war Asian nationalism.

Chapter 2

Planning for Trouble

Perhaps the least forgivable of all military blunders are those that are committed in the planning stage. For they often condemn the soldiers involved – officers and men alike – to certain defeat before the battle has even begun.

A number of different errors can be made: over-reliance on one element of a plan (the artillery bombardment at the Somme); inadequate reconnaissance (Bravo Two Zero); ignoring adverse intelligence (Arnhem); or just plain recklessness (the Jameson Raid, Colenso, and the Dieppe Raid). But they all have the same result: a gratuitous loss of life among the poor troops who have to put these lunatic schemes into action (of the instigators, Jameson is an exception in that he actually took part).

The Jameson Raid

Dr Jameson's invasion of the Transvaal with around 500 mounted men, on 29 January 1896, was one such occasion that was doomed to failure. It was part of a hare-brained scheme cooked up by Cecil Rhodes and British-born Johannesburg business leaders – with the tacit support of the British government – to replace President Paul Kruger's regime with one more sympathetic to their interests. But when other key elements of this unlikely plot were cancelled, Jameson chose to go ahead regardless – and suffered the inevitable consequences.

Born in Edinburgh on 9 February, 1853, Leander Starr Jameson was the youngest of his solicitor father's eleven children. Having qualified as a doctor in London in 1877, he looked set for a distinguished career until overwork threatened his health and he decided to join a practice in Kimberley, in the Cape Colony of South Africa. There he met Cecil Rhodes – the founder of the De Beers Mining Company – and the two struck up a close friendship.

Rhodes was a visionary who longed to see a federation of South African states under British rule. In particular, he wanted the British government to declare a protectorate over the lands to the north of the Transvaal and Bechuanaland. When, in 1889, it became clear that the Marquess of Salisbury's administration had no intention of increasing Britain's imperial responsibilities, he formed the British South Africa Company to develop the lands instead.

The following year he became Prime Minister of the Cape Colony and persuaded Jameson, at heart a gambler and a man of action, to give up his lucrative practice at Kimberley and travel north to establish the BSA Company in the tribal lands of the Matabele and Mashona. In September, Jameson raised the Union Flag on the site of modern Harare and became the company's administrator.

But the fearsome Matabele – a sub-tribe of the even more fearsome Zulus – were an obvious threat to company rule and had to be dealt with sooner or later. A pre-emptive war was duly launched by Jameson in 1893, partly to resuscitate the company, which was in debt and performing badly on the Stock Exchange. In January 1894, after a brief but brutal campaign, the Matabele chiefs submitted and Jameson's military self-belief knew no bounds. The new territory – more than 444,000 square miles in all – was named Rhodesia in honour of the company's founder.

But there was a far more serious obstacle to Rhodes' vision:

the Boer province of the Transvaal. In February 1881, at Majuba Hill, the Boers had defeated a British army and, under the terms of the ensuing peace treaty, had largely regained their independence. Five years later, gold was discovered on the Witwatersrand, close to Johannesburg, giving the Transvaal the means and the incentive to reject any proposals for federation under Britain's aegis.

By 1895, exasperated by the Transvaal's growing wealth, strength and independence, Rhodes was determined to overthrow Kruger's nationalist Afrikaner government. The opportunity was provided by the grievances of the *uitlanders* ('outlanders', i.e. non-Afrikaner immigrants), men of mainly British origin living and working in the Transvaal, particularly in the goldfields around Johannesburg. Initially, their chief complaint had been constitutional, in that they were denied the right to vote. This developed into a revolutionary movement because the largely British or non-Afrikaner mining interests of the Rand objected to those aspects of Kruger's policy that limited their profits – particularly high taxation and railway charges, the state dynamite monopoly, and his tariff policy which raised the cost of living and the level of wages.

As the owner of Consolidated Gold Fields, Rhodes had similar financial interests in the Transvaal. But his main reason for wanting to see the establishment of a more pro-British government there was that it would bring his dream of a South African federation closer to reality.

Rhodes's plan had three essential elements: an *uitlander* uprising would take place on the Rand; it would be supported by the rapid advance of an armed forced of BSA Company police, placed on the eastern frontier with Bechuanaland; and the British High Commissioner would then travel to Johannesburg so that he could conduct negotiations with the Boer government, discourage German intervention in support of

Afrikaner nationalism, and prevent the mine-owners from declaring their own independent republic.

The tacit support of the British government had been given in December 1894 when Rhodes confided his plan to Lord Rosebery, Prime Minister of the tottering Liberal government which had succeeded Salisbury's Tory administration. Rosebery's only stipulation was that the company police must not move until the rising in the Transvaal had actually begun. It was thanks to Rosebery's influence that Rhodes secured the appointment of Sir Hercules Robinson as High Commissioner for South Africa in early 1895. A former Governor of the Cape who had become a shareholder in the British South Africa Company and a director of De Beers, Robinson was a close friend of Rhodes's and a keen supporter of his plan.

In June 1895, the Liberal government fell and was replaced by Salisbury's Conservatives. But official support for Rhodes's plan remained. On 4 November 1895, Robinson sent a private memorandum to Joseph Chamberlain, the Colonial Secretary. Its main points were: a revolt would take place in the Transvaal 'sooner or later, and an accident might bring it about any day'; immediately on receiving news of the rising and the establishment of a provisional government, Robinson would 'issue a proclamation directing both parties to desist from hostilities and to submit to his arbitration'; the British government should 'notify their intention of supporting this attitude' and inform the press that 'a large force had been ordered to hold itself in readiness to proceed [from Britain] to South Africa'; he would at once go to Pretoria and order the establishment of a Constituent Assembly elected by every white male in the country.

'Agree generally with your idea,' Chamberlain replied on 6 December. 'I take for granted that no movement will take place unless success is certain. A fiasco would be most disastrous.'

Meanwhile, the rest of the plan was taking shape. By the end of November, Jameson had gathered about 600 men near to Transvaal's eastern border: 350 were at Pitsani in Bechuanaland, mostly from the Company's Mashonaland Mounted Police; the rest had been recruited from the Bechuanaland Border Police and were at Mafeking in the northern Cape Province. Besides rifles, the force was armed with six Maxim machine-guns and two field guns. In addition, the Rhodesia Horse, a 1,000-strong volunteer force set to invade the Transvaal from the north, had been put in readiness near the frontier under Captains Spreckley and Napier. Their commander, Colonel Willoughby, had joined Jameson as chief of staff.

At the same time, 5,000 rifles, three Maxims and 1,000,000 rounds of ammunition were in the process of being smuggled into Johannesburg. The intention was to supplement them with arms from the Boer arsenal at Pretoria, which would be seized as soon as the *uitlander* rising began. Needless to say, Rhodes picked up the tab for all these preparations – to the tune of £61,500.

On 19 November, Jameson met the chief plotters in Johannesburg. The provisional date for the rising was fixed for midnight on 28 December. But Jameson refused to enter the Transvaal 'like a Brigand' and, reluctantly, the *uitlander* leaders gave him a letter formally asking him to come to the aid of the people of Johannesburg. It was also agreed that, as soon as trouble started, Sir Hercules Robinson would leave for the Transvaal to arbitrate a settlement.

But marked differences arose about the future of the Transvaal. Rhodes did not want to commit himself, and the *uitlanders* were rightly suspicious of British interference. Most of them were keen to establish a liberal republic which would accommodate the needs of the mining community; they did not want to become part of a British colony.

As the day of the rising approached, enthusiasm for it in Johannesburg diminished. The arms that had been smuggled in – mostly rifles – were fewer than promised; the plan to seize the Pretoria arsenal had to be abandoned for lack of recruits. By 26 December, the divided *uitlander* leaders had decided to postpone the rebellion at least until 6 January.

Jameson, who had been informed on 24 December that the operation would go ahead as planned, was now told of the hold-up as he waited impatiently at Pitsani on the Bechuanaland-Transvaal border. His response, on 27 December, was to point out that some of his police had already gone forward to carry out 'distant wire cutting' and that if they could not be stopped he would have to abide by the original plan.

The following day, in a stream of coded telegrams from the British South Africa Company's head office in Cape Town, Jameson was told in no uncertain terms that the Johannesburg plot had collapsed: '. . . public will not subscribe one penny towards it even with you as director.' He was also told that a Major Heaney was on his way with a message from the *uitlander* leaders. Once Jameson had passed on what Heaney had to say, they could make a joint decision about what to do: 'you and we must judge regarding flotation.'

'It is all right if you will only wait,' his instructions ended. But 'we cannot have fiasco.'

Jameson's reply was typically obstinate: 'Unless I hear definitely to the contrary shall leave tomorrow evening.'

He sent it at 5pm, probably with the knowledge that it would not reach Cape Town until long after the company offices had closed for the weekend. In the event, Rhodes did not receive it until the following day.

It was only during the evening of Saturday, 28 December, that Rhodes definitely decide to cancel the operation after hearing from *uitlander* leaders that not even a token rising would take place. It had not been possible to raise a revolutionary force and, in any case, 'the Boers had got on guard'.

Robinson, the High Commissioner, was informed that 'the whole thing has fizzled out' the following morning. But no cancellation message was sent to Jameson.

Heaney reached Pitsani at about 10.30am on the 29th, and told Jameson the Johannesburg leaders wanted him to postpone his expedition because they lacked arms and had quarrelled about the future of the Transvaal. After walking up and down outside his tent for some time, mulling over this news, Jameson re-entered and announced that he was going into the Transvaal. Heaney agreed to accompany him.

Part of the reason for Jameson's hasty decision was a Reuters report, published in the South African press the day before, of rumours in Johannesburg about the 'secret arming of miners and warlike preparations'. Women and children were said to be leaving the town, while Kruger and Piet Joubert, the Commandant-General of the Transvaal's forces, had returned to Pretoria and the leading conspirators were hoping to come to an arrangement with them. With the cat almost out of the bag, Jameson was determined to move before Kruger could take the necessary countermeasures. An additional incentive was his knowledge of what a postponement would do to the already depressed finances of the company.

His final message to Rhodes, sent on Sunday morning, stated that he would 'leave to-night for the Transvaal'. It concluded: 'We are simply going in to protect everybody while they change the present dishonest government and take vote from the whole country as to form government required by the whole.'

It was Jameson, therefore, who alone took the decision to go into the Transvaal to force a rising, despite the fact that he knew the *uitlanders* were unenthusiastic and Kruger's government had got wind of the plot. Such an enterprise was doomed before it had begun.

With his office closed for the weekend, Rhodes did not receive Jameson's final message, along with the one sent the previous day, until 1pm on the 29th. His response was to forbid the expedition: 'Things in Johannesburg I yet hope to see amicably settled, and a little patience and common sense is only necessary – on no account must you move, I most strongly object to such a course.' Unfortunately, it was a Sunday and the telegraph offices closed at lunchtime. This vital message could not be sent until the following day – and by then it was too late.

On Sunday afternoon, Jameson announced to his men that they were about to ride into the Transvaal at the request of the citizens of Johannesburg. It would probably be a bloodless expedition, he said, and everyone would be paid a bonus for special service.

That night, having cut the telegraph wires, they crossed the border. The Mafeking contingent left at about the same time, also having severed communications, and the two columns met up at Ottoshoop in the Transvaal at five the following morning. The total force numbered 510 mounted men with eight Maxims and three field guns. Carrying their stores and ammunition were eleven mule-drawn carts and thirty pack horses, tended by seventy-five natives.

Although Jameson had the wires to Zeerust cut before leaving Ottoshoop, a rider was sent there by a local official to warn Louis Botha, the district commandant. The wires between Zeerust and Pretoria should already have been down but the men responsible failed to carry out their task. One story relates how they got drunk and cut fencing wire instead. In any case, Botha was able to send Joubert details of the invasion. Within four hours of their departure from Ottoshoop, the Commandant-General knew the direction they were riding in, their strength and what guns they had.

Early the following morning, more than 90 miles from Pitsani, a lone rider caught up with the column. He had been sent to inform Jameson and his officers that the High Commissioner had repudiated their action and ordered them to return. They took no notice. Next morning, New Year's Day 1896, a second rider reached them with a similar message. Jameson replied that, much as he would have liked to obey the order to retire, he could not because the supplies to his rear were exhausted and he was anxious 'to fulfil my promise . . . to come to the aid of my fellow men in their extremity'.

They pressed on, but by now the column was in poor shape. The food and forage arrangements had not proved adequate, and most of the remounts collected at a secret rendezvous had proved to be carthorses not used to the saddle. Armed Boers were in evidence, and Jameson allowed his men very little rest.

That afternoon, having advanced an impressive 154 miles in three days, the column was intercepted by Boers near Krugersdorp, just 20 miles from Johannesburg. Small groups had been shadowing their flanks and rear ever since they had crossed the border; but the commandos (the name originally given to forces of armed Boer volunteers) had not been able to gather in sufficient force to oppose them until now.

Numbering 550 men under Commandants Cronje, Malan and Potgieter, they had taken up strong fortified positions on a rise about 3 miles west of the town. Jameson spotted them as he reached the top of a steep incline at the other side of the valley. He resolved to blow them out of their positions and ordered the field guns to open up. Fortunately for the Boers, they were well entrenched and the shrapnel shells only wounded a few horses.

Deceived by the complete silence from the Boer lines, Jameson now sent a party of 100 horsemen, supported by two Maxims, to investigate. Advancing in extended line, they were crossing a muddy stream, within 700 yards of the ridge, when the Boers opened fire from front and flank. About thirty fell dead or wounded and another thirty crept into nearby reeds for cover and were later captured. The rest withdrew.

With the road blocked, Jameson decided to try and reach Johannesburg by skirting to the south in the direction of Randfontein. The field guns and machine-guns provided covering fire, blowing up a disused battery-house held by the Boers in the process.

At first, all seemed to be going well as, according to Willoughby, they 'obtained' a guide who put them on 'a road leading direct to Johannesburg'. In fact, the guide was a Boer scout who led them straight into an enemy trap. Surrounded on three sides by Boers, Jameson laagered the wagons and prepared to fight it out. But no attack was made that night, although intermittent firing prevented the men from sleeping; they had not eaten for eighteen hours.

At dawn on 2 January 1896, the column struck camp and moved south, the only direction not commanded by the Boers. As before, its rear and flanks were protected by the guns. The Boers followed. Having gone about 10 miles, the column reached Vlakfontein farm and found more Boers blocking their way. Many lay along a low cliff to the right, others were on a hillock called Doornkop further on. It was impossible to skirt the cliff, and equally impossible to retreat.

Exhausted, surrounded and heavily outnumbered, Jameson's column prepared to make its last stand. From an outbuilding of the farm they began to bombard the cliff until the machine-guns overheated and jammed for lack of water and the field guns ran out of ammunition. Freed from this heavy fire, the Boers poured well-aimed rifle bullets into the defensive perimeter, taking a heavy toll of men and horses. Then the Boers' Staats Artillerie (effectively, the only Regular unit of the Boer forces) came into action on the left, and it was soon all over.

Someone tied a white apron to a stick and hoisted it over the outhouse. Spotting it from his position on the Doornkop, Cronje sent a message to Jameson asking him what his intentions were. But Willoughby had already sent off a note to 'the Commandant of the Transvaal Forces', offering to surrender 'provided that you guarantee a safe conduct out of the country for every member of the force'.

Cronje's note, in response, was that they would be spared if they gave up their flags and arms, and agreed to pay for the expenses they had caused the Transvaal. The Boer commandant then rode up to the farmhouse, followed by his burghers. Inside, he found Jameson and his officers looking dirty and miserable. Some stood round a badly wounded man and wept. Jameson, trembling 'like a reed', told Cronje that he fought under no flag but was ready to lay down his arms.

At this moment, Malan arrived and told Cronje that he had no authority to arrange terms. It had to be unconditional surrender or nothing. When told of this, Jameson took off his hat, bowed and said: 'I accept your terms.'

Jameson and his officers were taken to Pretoria in carts; his men on horseback. Their casualties numbered seventeen killed and fifty-five wounded. The Boers, by contrast, had lost just four killed – one shot by his own side and one by a wounded trooper he had gone to help – and three wounded. But the most serious casualty was Cecil Rhodes, who resigned as Prime Minister of the Cape on the day he learnt that Jameson had surrendered. 'Poor old Jameson,' he exclaimed. 'Twenty years we have been friends, and now he goes in and ruins me.'

So ended the ill-fated expedition that came to be known as the Jameson Raid. While it had been part of a plot that included an uprising in Johannesburg and the involvement of the British High Commissioner, it had had some hope of success – however minimal. But when both of these prerequisites fell through, Jameson's poorly planned mission was bound to end in disaster.

The Kruger Telegram

No sooner had the news of Jameson's raid reached Germany than the Kaiser, Wilhelm II, bristling with martial rage, was suggesting a German 'protectorate' over the Transvaal and the dispatch of troops to South Africa. Having managed to quash these reckless proposals, his ministers decided to placate their imperial master by agreeing to send President Kruger a telegram of congratulation.

'I express to you my sincere congratulations,' it read, 'that you and your people, without appealing to the help of friendly powers, have succeeded . . . in restoring peace and in maintaining the independence of the country against attack from without.'

It contained two innuendoes that were meant to irritate the British government. Mention of the 'independence' of the Transvaal ignored the limitations imposed on its autonomy by the London Convention of 1884 (one of the agreements which followed the First Boer War), and the reference to 'friendly powers' implied that Germany might have intervened if asked.

But it was certainly not intended to provoke the storm of anti-German feeling with which the British public and press greeted its publication. Facing a very real possibility of war, the German government changed tack and withdrew its support for a conference to guarantee the future integrity of the Transvaal. Kruger was left to make what terms he could in direct negotiation with the British government.

The telegram had one other major consequence: it encouraged the British public to condone the failure of the raid. Alfred Austin, the Poet Laureate, even went so far as to compose a laudatory piece entitled 'Jameson's Ride' for publication in *The Times*. The following verse was typical:

> There are girls in the gold-reef city.
> There are others and children too!
> And they cry, 'Hurry up! for pity!'
> So what can a brave man do?

Jameson's subsequent rehabilitation was nothing short of extraordinary. Returned to London to stand trial for an offence against the Foreign Enlistment Act, he was sentenced to fifteen months' imprisonment in July 1896. But with his health rapidly deteriorating in Holloway Prison, he was released after serving just one-third of his sentence. Present throughout the Siege of Ladysmith during the Second Boer War, he was elected a member of the Cape Parliament in June 1900. When Rhodes died, in 1902, he succeeded him as leader of the Progressive Party. Two years later, and just eight since his conviction, the Progressives gained a majority and he became Prime Minister of the Cape Province. A baronetcy followed in 1911, six years before his death.

Colenso

The Battle of Colenso on 15 December 1899, during the Second Boer War, was the third of three major British defeats in what came to be known as 'Black Week'. But Colenso stands alone as the classic example of a battle that could not be won, a battle that should never have been fought.

Six weeks earlier, the British had suffered their first serious reverse of the war at Nicholson's Nek near Ladysmith. Within days the Boers had cut the railway and telegraph lines to the south of the town, trapping 12,000 troops, many more civilians and a vast quantity of stores. Also surrounded by this time were the major population centres of Mafeking and Kimberley in the northern Cape.

When General Sir Redvers Buller, the commander of the 50,000-strong reinforcement Army Corps, dispatched to South Africa on the outbreak of the war between Britain and the 'rebellious' Transvaal and Orange Free State, arrived in Cape Town by sea on 31 October 1899, it was not a moment too soon. 'There is no stronger commander in the British Army,' wrote a contemporary, 'than this remote, almost grimly resolute, completely independent, utterly fearless, steadfast, and always vigorous man.'

The holder of the Victoria Cross (which he had won in the Zulu War of 1879), Buller was very much a man of action. Born in 1839, the grand-nephew of the Duke of Norfolk, he had been expelled from Harrow before joining the 60th Rifles at the age of nineteen. Since then he had served with distinction in no less than five campaigns, including the Zulu War of 1879, during which he had commanded the irregular cavalry. But though personally fearless and a first-rate subordinate, he had never held a major independent command in war, and would prove himself incapable of making the right decision under pressure. In the words of one historian: 'He made a superb major, a mediocre colonel, and an abysmally poor general.'

At the time, the British press had no such reservations. Despite rumours that he was an alcoholic, they hailed his appointment as a masterstroke. 'The Boers'll cop it now,' said one urchin to another in a *Punch* cartoon. 'Farfer's gone to South Africa, *an' tooken 'is strap*!'

Buller's instructions from the War Office were for the Army Corps to strike north from the Cape Midlands in the direction of the two Boer capitals – Bloemfontein in the Orange Free State and Pretoria in the Transvaal. With Ladysmith in danger of capitulating, however, Buller decided to ignore these instructions and split the Army Corps into three unequal parts. He would relieve Ladysmith with the largest; Lieutenant-General Lord Methuen would march on Kimberley and Mafeking with the next biggest; while Lieutenant-General Sir William Gatacre would use the rest to contain the Boer invasion of the Cape Province until Buller could reassemble the Army Corps to carry out the original plan.

Aware that the British would soon attempt to relieve Ladysmith, the Boer commander, Commandant-General Joubert, detached 4,000 men to intercept them. Led by the young and vigorous Louis Botha, by now a general, this force quickly overran Colenso on the south bank of the Tugela River, beyond the hills that fringed the Ladysmith plain. By 20 November – having already captured part of an armoured train containing a youthful Winston Churchill, war correspondent for the *Morning Post* – the Boers had encircled Estcourt and their patrols had advanced within 10 miles of Durban.

Two days later, Buller arrived and began to assemble his force of 20,000 men. With the Tugela River coming down in flood, there was a danger that the Boer force would be cut off, so Joubert ordered it to retire north. Crossing the river at Colenso, they blew up the road and railway bridge and, reinforced by men from the Ladysmith siege lines to a total strength of 8,000 men, began to entrench on the north bank.

At dawn on 12 December, Buller set out with 18,000 men from his advance base at Frere, 10 miles south of Colenso. His intention was to bypass the strong Boer position there which was, in his opinion, impregnable. He would avoid it by crossing the river at Potgieter's Drift further upstream. But that same day he received the appalling news that Lord Methuen's army of 13,000 men had been routed at Magersfontein on 11 December, having already suffered a serious reverse at Modder River on 28 November, as though General Gatacre's defeat at Stormberg Junction, on 10 December, had not been bad enough. Clearly, Buller's presence was urgently required in the Cape. But he could not bring himself to abandon Ladysmith. Indecisive, he dithered for two days – while his long-range artillery shelled Boer positions – before making up his mind. He would attack Colenso head on and relieve Ladysmith by the shortest route.

It was a disastrous compromise. In anticipation of such a move, Botha had entrenched himself in the high ground overlooking Colenso in camouflaged lines that extended for 10 miles. Only on Hlangwane Hill was he vulnerable, for there his men were cut off from their comrades by a northerly sweep of the Tugela.

Buller's unimaginative plan of attack was for two brigades to advance against the Boer centre and right, and hopefully force the river, while his mounted brigade assaulted Hlangwane. Essentially a head-on attack against well-prepared and unlocated enemy positions, it was doomed to failure. Not least because Buller, fearful of spies, had failed to inform Lieutenant-General Sir George White, the senior officer at Ladysmith, who was planning a break-out to coincide with the attack.

At 5.30am on 15 December, Buller's heavy guns opened up on the suspected Boer positions at a range of three miles. In the centre, Major-General Henry Hildyard's 2 Brigade began to advance, supported by twelve field and six naval guns commanded by Colonel Charles Long, famous for leading

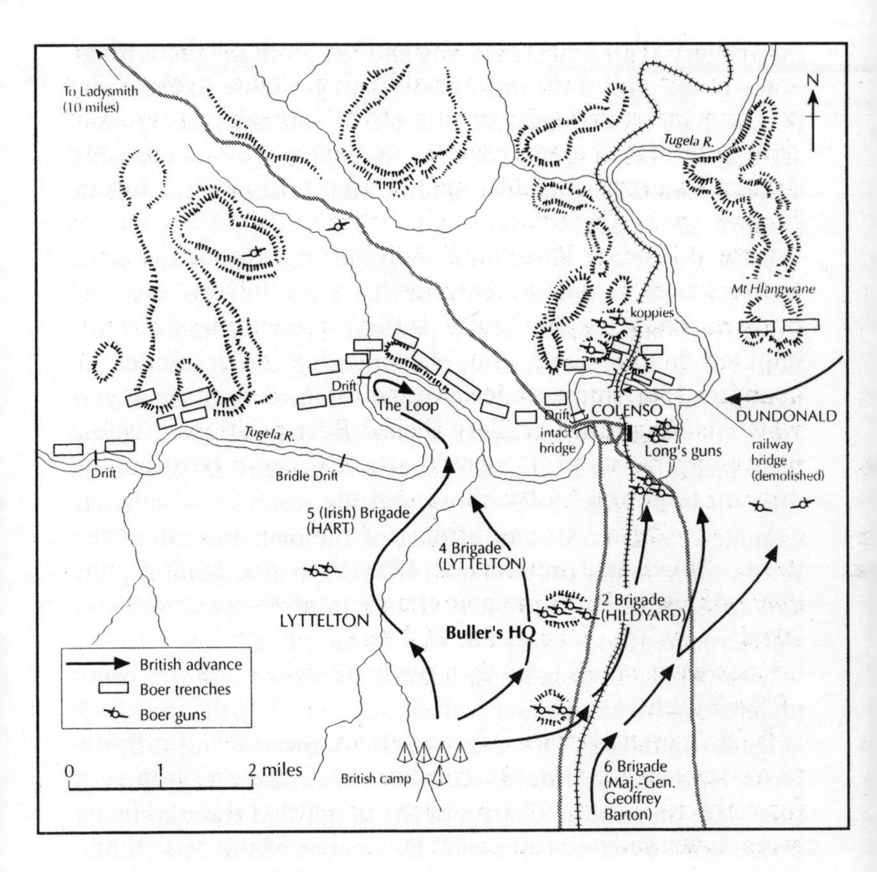
To Ladysmith
(10 miles)
N
Tugela R.
Mt Hlangwane
koppies
Drift
The Loop
Drift
COLENSO
DUNDONALD
intact
bridge
Long's guns
railway
bridge
(demolished)
Tugela R.
Drift
Bridle Drift
5 (Irish) Brigade
(HART)
4 Brigade
(LYTTELTON)
2 Brigade
(HILDYARD)
LYTTELTON
Buller's HQ
British advance
Boer trenches
Boer guns
0
1
2 miles
British camp
6 Brigade
(Maj.-Gen.
Geoffrey
Barton)

the artillery at Omdurman. Long had been told to use his long-range naval guns until the infantry had got close to the river, but he had other ideas. A disciple of the new theory that artillery should be used to destroy the enemy at close range, his horse-drawn field guns forged ahead until they were a mile in front of the infantry.

With the Boers silent and invisible, the field guns were within a few hundred yards of the river, to the right of Colenso village, when Long ordered them to halt for action. At that moment, a single shot rang out from the far bank, quickly followed by a storm of fire from rifles and artillery. Caught in the open, struggling desperately to bring their pieces into action, the gunners were easy targets. Less than an hour later, with Long and his second-in-command wounded, and more than a third of the men casualties, the acting commander ordered the survivors to abandon the guns and take shelter in a nearby donga (a steep-sided gully or watercourse, dried-up except during the rains).

Meanwhile, an even more calamitous manoeuvre was being undertaken by Major-General Fitzroy Hart's Irish Brigade (5 Brigade) on the left. Ordered to make for the crossing at Bridle Drift, Hart set off with his battalions in close order as though they were on parade. Fearful of the target that these compact bodies would make, Colonel C.D. Cooper of the 2nd Dublin Fusiliers tried to open out his battalion to twice the parade-ground interval. Hart overruled him. Noted for his preoccupation with drill – the brigade had been subjected to half an hour of it before moving off – he liked to keep his men 'well in hand'. Even when the cavalry, screening his left flank, warned him that Boer entrenchments lay directly ahead, he saw no reason to alter his formation.

By 6.15 am he had reached a point some 300 yards from the beginning of a large northerly loop in the river, forming a salient hemmed in on three sides by water. Stopping to consult his map, he saw that Bridle Drift was some way to the left of the loop. Unfortunately, his native guide insisted that the drift, the *only* crossing place, was in the loop itself. Instead of sending out patrols, Hart took the native at his word and ordered his men into the salient.

Moments later, the Boers opened fire. Instinctively, some of his men began to deploy and open fire. But Hart, anxious to reach the ford despite the danger, urged them forward. As they advanced, their packed ranks fired on from three sides, all order soon disappeared. Small groups, led by officers, would run forward, 50 yards at a time, before taking cover again. 'All pushed forward blindly,' wrote one officer, 'animated by the sole idea of reaching the river bank.'

By chance, one of Hart's battalions, the 1st Royal Inniskilling Fusiliers, had strayed to the left of the loop and were soon nearing the real Bridle Drift. Part of it had been dug away by the Boers, but it was probably still fordable. But Hart then ruined any chance of salvaging something from the disaster by ordering the Inniskillings into the loop.

Within 40 minutes, more than 400 men had been killed or wounded. Buller – watching the fiasco from the heavy gun positions on high ground to the rear – had already dispatched two gallopers to warn Hart to keep out of the loop. Now he sent a third, his Military Secretary, Colonel the Hon. Frederick Stopford (see p. 50), and it finally had the desired effect. Covered by Major-General the Hon. Neville Lyttelton's reserve brigade (4 Brigade), Hart's battered remnants retired (although many unwounded men remained pinned down and were later taken prisoner). It was 7am; Long's guns fell silent at about the same time.

By now, Lord Dundonald's mounted brigade had begun the attack on Hlangwane. Dismounting at the edge of a mealie field, they crept towards the slopes of the hill but were soon pinned down by accurate rifle fire. Realizing he could not advance without infantry support, Dundonald sent a message to the nearest of the two reserve brigades. Its commander refused to help. He had just received an order from Buller to stay out of trouble.

Buller, meanwhile, had ridden forward to see what he could do about extricating Long. Meeting Hildyard, who was about to launch his attack on the wagon bridge at Colenso and the two fords either side, Buller told him: 'I'm afraid Long's guns have got into a terrible mess. I doubt whether we shall be able to attack Colenso today.'

It was 8am. With only one of his four infantry brigades having seen action, Buller decided to call off the whole operation. His priority now was to rescue Long's twelve field guns, almost half his total artillery. Hildyard's task was to seize Colenso village to cover the guns from the right. This his men achieved despite the fact that, like Hart's brigade, they had to advance into a salient and were fired on from three sides. It helped that Hildyard deployed them in open order, with 6 yards between each man and 60 yards between each half-company.

Yet still the guns were stranded in open ground more than 500 yards away. Eventually Buller called for volunteers. Up stepped one corporal, six men and three officers (Captains Schofield and Congreve, and Lieutenant Freddy Roberts, the only son of Field Marshal Lord Roberts, VC, soon to supersede Buller as Commander-in-Chief). As the party galloped towards the guns with two limbers, the Boers opened up with everything they had. Roberts and Congreve were wounded and unhorsed. The rest, led by Schofield, whose clothes were ripped by six bullets, managed to hitch up two guns and escape.

Three further attempts were made; but not one reached the guns. One party had twelve horses shot, one man killed and five wounded. In all, five Victoria Crosses and eighteen

Distinguished Conduct Medals were awarded for this conspicuous heroism.

The only hope of saving the guns now was to leave Hildyard's infantry in position until nightfall – meaning eighteen hours without water. Yet Dundonald had been driven back from Hlangwane with heavy losses, leaving the British right flank dangerously exposed. Buller's fear was that he would lose much of his infantry as well as the ten guns if he remained at Colenso. He decided, therefore, to cut his losses and order his whole force to retire. It was barely midday.

This meant abandoning the wounded in exposed positions. 'No water, not a breath of air, and not a particle of shade and a sun which I have never felt hotter even in India,' recalled Captain Congreve, lying near to the badly injured Freddy Roberts. Found by the Boers during the afternoon, they and the other wounded were allowed to return to the British lines for treatment. Roberts died the following day.

The Boers could afford to be magnanimous. 'Today,' Louis Botha telegraphed Pretoria on the 15th, 'the God of our fathers has given us a great victory.' At a cost of just 40 men killed and wounded, he had captured 10 guns, 600 shells and inflicted more than 1,100 casualties (143 killed, 755 wounded and the rest taken prisoner).

Buller's official report to the War Office was that of a broken man: 'My failure today raises a serious question. I do not think I am now strong enough to relieve White [in Ladysmith]. Colenso is a fortress, which I think, if not taken on a rush, could only be taken by a siege . . . My view is that I ought to let Ladysmith go, and occupy good positions for the defence of South Natal, and let time help us . . .'

Early next day, he signalled to White at Ladysmith that it would take at least a month to wear down the Boer defences. 'Can you last so long? If not, how many days can you give me to take up defensive position, after which I suggest your firing away as much ammunition as you can, and making the best terms you can.'

Privately, Buller blamed Long for his defeat. 'I was sold by a damned gunner,' he later told a friend. In truth, Buller alone was responsible. Long's conduct, and Hart's for that matter, simply hastened the inevitable.

Buller should never have attacked Colenso in the first place. He knew that it was the strongpoint of the Boer defensive line, and yet he abandoned his original plan to circumvent it on the ludicrous grounds that Methuen's defeat meant his presence was urgently required in Cape Province. If that really had been the case, he should have abandoned the relief of Ladysmith altogether. Instead, he made the worst possible compromise by choosing the quickest route to the besieged town through a virtually impregnable Boer position.

He admitted as much in a letter to his wife: 'I had to have a play at Colenso, but I did not think I could get in . . . One knows the worst at any rate – I think quite between you and me that I was lucky in not getting in as if I had I should not have known what next to do [because the worst fighting was still ahead] . . . Better luck next time.'

Within a week of the battle, he had been superseded as overall commander by 67-year-old Lord Roberts. He stayed on as commander of British forces in Natal with the task of completing the relief of Ladysmith. This he finally managed in late February 1900, but only after two more humiliating defeats at Spion Kop (see p.35) and Vaalkrantz. In October, having salvaged some of his miliary reputation with a string of modest victories, he returned to Britain to resume his command of the Army Corps at Aldershot. But his earlier blunders had not been forgotten and a year later he was sacked by Roberts, the new Commander-in-Chief, who had never forgiven him for the death of his son.

The First Day on the Somme

The First of July 1916 – the opening day of the Battle of the Somme – has been described by one military historian as 'the blackest day of slaughter in the history of the British Army'. Half the attacking force of 120,000 became casualties, 1 in 6 was killed. The fact that the final plan had little chance of achieving its objectives makes the death toll even less forgivable.

The die had been cast at the Chantilly Conference in early December 1915 when the four Great Powers at war with Germany and Austria–Hungary – Britain, France, Russia and Italy – had agreed to launch simultaneous offensives in the coming year on three fronts – Western, Eastern and Italian. It was a personal victory for General Joseph-Jacques Joffre, the French Commander-in-Chief, who had dismissed the Gallipoli campaign of 1915 as a sideshow diverting valuable resources from the main effort in the west.

Later that month, Joffre had a private meeting with General Sir Douglas Haig, the new British Commander-in-Chief, to discuss the offensive. Haig's preference was for the British to attack in Flanders so that the German line could be rolled up from the north. But with the instructions of the Supreme War Council ringing in his ears – that 'the closest co-operation between the French and the British as a united Army must be the governing policy' – he was forced to yield to Joffre's demand that the attack be a joint effort at the point where the British and French lines joined: along the River Somme.

Already the Allies had a numerical superiority on the Western Front with 139 divisions (38 British, 95 French and 6 Belgian) to Germany's 117. But with the British Army rapidly expanding, the offensive was tentatively set for August when a further nineteen of Kitchener's New Army divisions would be available. Initially, twenty-five British divisions were earmarked to attack north of the Somme, forty French divisions south of it.

All this changed when the Germans launched an offensive against the French at Verdun in February 1916. Over the next few months, a total of 115 divisions were fed by both sides into the 5-mile front. By the time the fighting ended in June, the French had lost 315,000 men, the Germans 281,000. Of greater significance for the British was the fact that Haig, in response to Joffre's appeal for help, had agreed not only to extend the British share of the front line but also to put forward the date of the Somme offensive to late June.

These changes undermined Haig's confidence in the coming battle. His raw troops, with less time to train, would now shoulder the burden of the attack with 13 divisions advancing on a front of 17 miles. The veteran French, on the other hand, would move forward with just 11 divisions on a front of 8 miles (as opposed to the 25 miles previously agreed). Victory in 1916 was no longer possible, Haig told his army commanders in May. The offensive now had three more limited aims: to relieve pressure on the French at Verdun by drawing off German divisions to meet the British attack; to sap the strength of the Germans; and to put the British Army in a position from which it could strike the final blow in 1917.

The main part of the attack was to be carried out by Lieutenant-General Sir Henry Rawlinson's Fourth Army, the formation immediately to the north of the French. Its principal assault would be by 9 divisions on a 10-mile front from Montauban in the south up to the River Ancre, a tributary of the Somme. At the same time, 3 more of its divisions would secure its left flank on a 3-mile front north of the Ancre. A little further to the north, and partly as a diversion, two of Lieutenant-General Sir Edmund Allenby's Third Army divisions would attempt to eliminate the awkward German salient at Gommecourt. The French, meanwhile, would attack in the sector to the south of Montauban.

If all went well, Lieutenant-General Sir Hubert Gough's Reserve Army (which was redesignated Fifth Army during the battle) of 3 cavalry divisions would push through the gap and take the town of Bapaume, 9 miles behind enemy lines, before turning north into open country. Rawlinson's infantry would roll up the German trench line by heading in the same direction, while the French would secure the right flank of the breakthrough and prevent intervention by German reserves from the south.

Having outlined the main objectives, Haig allowed his army commanders to fill in the details. For General Rawlinson the key to success was artillery. Impressed by the initial German gains at Verdun, when prolonged shelling had destroyed the French front-line trenches, he intended to repeat the trick on the Somme. Using every gun available, he planned to bombard the German lines for five days and nights before the attack began. This would, it was hoped, blast gaps in the barbed wire protecting the German lines, destroy the trenches, entomb many of the defenders in their dugouts and disrupt communications to the rear. Then, as the infantry advanced, the artillery would put down a barrage or 'curtain of fire' ahead of them to prevent the enemy from reoccupying their defensive positions. In theory, by a number of carefully timed 'lifts', this 'creeping barrage' would enable the infantry simply to stroll across and take possession of the enemy trench system.

So that the infantry would not get ahead of the artillery, it was decided that they would go 'over the top' in a series of waves, at one-minute intervals, moving at a methodical pace of no more than 100 yards in every 2 minutes (less than 2 miles per hour). As every man would be carrying at least 70 pounds of equipment, to avoid exhaustion they were instructed not to run until within 20 yards of the enemy trench. 'Assurance was given that the time-honoured system of short rushes would, in this instance, be unnecessary,' recalled a corporal of the 9th York and Lancaster Regiment.

Rawlinson's plan was to take one enemy trench line at a time. Once the inevitable counter-attacks had been repelled, his artillery would bombard the second line until it, too, was ready for capture; then it would move on to the reserve line of trenches.

It was eventually decided that the attack would go in at 7.30am on 29 June. The established principle of attacking at dawn – more than three hours earlier – had been sacrificed in favour of better artillery observation. Taking part in the attack would be eighteen infantry divisions, each containing thirteen battalions (one of them a Pioneer battalion), as well as divisional troops such as artillery, engineers, medical and transport services, and so on. The Germans, by contrast, had just six weak divisions defending the British sector. However, more than 60 per cent of the attacking troops were New Army men yet to receive their baptism of fire.

When Rawlinson eventually submitted his plan for approval, Haig was disappointed. Convinced that it would not lead to a breakthrough in which cavalry could be deployed (Haig had been a cavalry commander in the Boer War), he made three suggestions: to shorten the bombardment to give the Germans less warning of an attack; to rush the trenches as soon as the barrage on them lifted; and to take at least two of the German lines on the first day.

Rawlinson rejected the first two outright. He felt the lack of heavy artillery meant a lengthy bombardment was necessary to destroy the German defences, while the troops involved were insufficiently trained to carry out more sophisticated infantry tactics. Only on the third point was he prepared to compromise. North of the main road to Bapaume, in what was roughly half the sector due to be attacked, the German second line lay relatively close to the first. Here, at least, he agreed to make the support line a first-day objective because it was not beyond the effective range of his artillery (4,000 yards ahead of the forward British trench). Further south, where the gap between the first two German lines was greater, the guns would have to be dragged forward before the second line could be taken.

By the early hours of 24 June, when the great bombardment opened, the British had 1,500 field guns, howitzers and mortars in position – one for every 17 yards of the enemy front. More shells were fired over the next week – 1,508,652 – than had been used in the whole of the first year of war. Each morning, between 6.25 and 7.45am, every gun delivered as concentrated a fire as possible. The plan was to shorten this programme by 15 minutes on the day of the attack, so that when the infantry attacked at 7.30am the Germans would be expecting more shelling. A constant, but less violent, bombardment continued for the rest of each day, with half the guns silent at night.

For a number of reasons, the bombardment was hopelessly inadequate. 'The whole thing depended on our artillery being able first of all to locate and then smash up the concrete machine-gun posts and then with the field guns to sweep away the wire entanglements,' wrote Captain (later General Sir) James Marshall-Cornwall, Haig's Intelligence Officer. 'But, unfortunately, the weather broke. For five days out of the six of the bombardment there was low cloud and drizzle. Air observation was impossible and artillery observation was very hampered. The fact was that neither did they pinpoint the machine-gun posts opposite them, they also failed to cut the wire.'

The method of cutting barbed wire was to use a shrapnel shell which burst into a hail of steel balls about 20 feet above

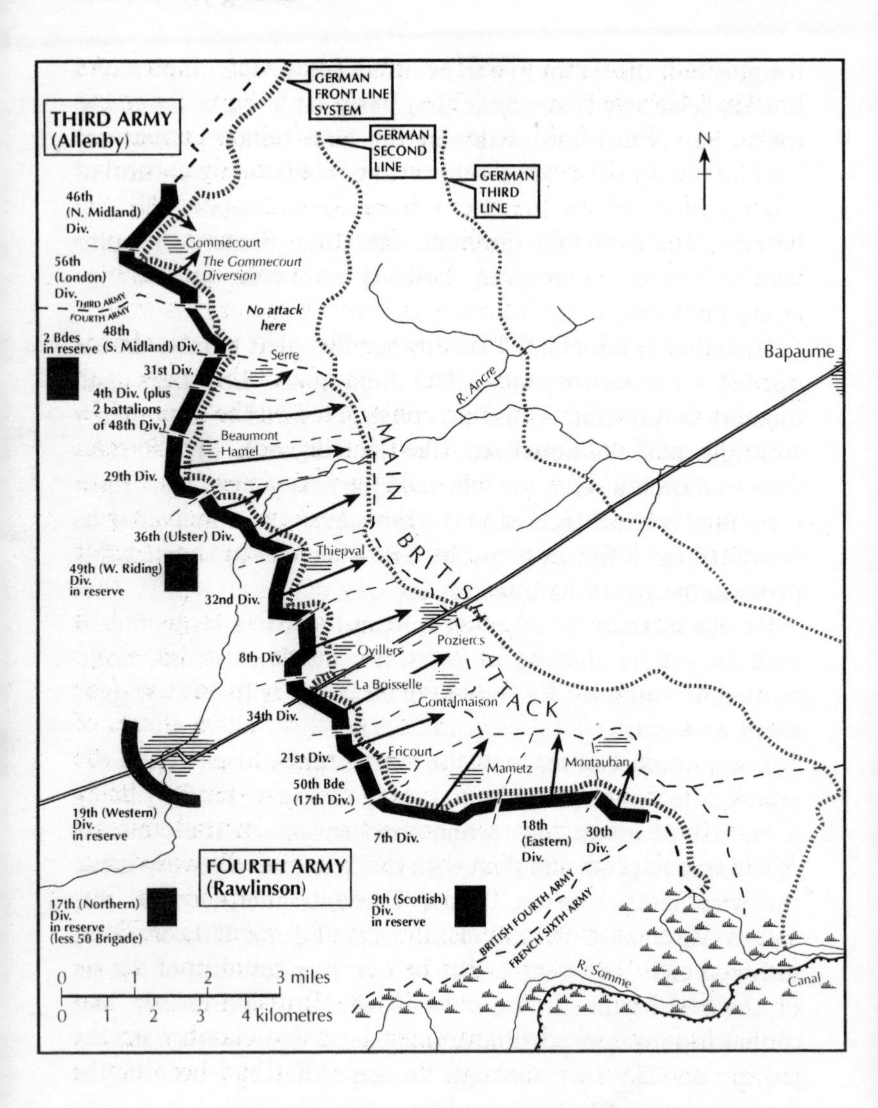
GERMAN
FRONT LINE
SYSTEM
GERMAN
SECOND
LINE
GERMAN
THIRD
LINE
N
THIRD ARMY
(Allenby)
46th
(N. Midland)
Div.
Gommecourt
56th
(London)
Div.
The Gommecourt
Diversion
THIRD ARMY
FOURTH ARMY
No attack
here
2 Bdes
in reserve
48th
(S. Midland) Div.
Serre
31st Div.
4th Div. (plus
2 battalions
of 48th Div.)
Beaumont
Hamel
29th Div.
R. Ancre
Bapaume
MAIN BRITISH ATTACK
36th (Ulster) Div.
49th (W. Riding)
Div.
in reserve
Thiepval
32nd Div.
8th Div.
Ovillers
Pozieres
La Boisselle
Contalmaison
34th Div.
21st Div.
Fricourt
Mametz
Montauban
50th Bde
(17th Div.)
19th (Western)
Div.
in reserve
7th Div.
18th
(Eastern)
Div.
30th
Div.
FOURTH ARMY
(Rawlinson)
17th (Northern)
Div.
in reserve
(less 50 Brigade)
9th (Scottish)
Div.
in reserve
BRITISH FOURTH ARMY
FRENCH SIXTH ARMY
R. Somme
Canal
0 1 2 3 miles
0 1 2 3 4 kilometres

the ground. But many of the time fuses were faulty. As Marshall-Cornwall noted: 'They didn't all burn the right length and, I'm afraid, a lot of the half-trained gunners of the New Army divisions didn't set the fuses exactly accurate.'

Two-thirds of all the shells fired were shrapnel. Of the others (roughly half a million), less than 8 per cent were high-explosive. There was, besides, an acute shortage of heavy guns and ammunition, with only 34 pieces of a calibre greater than 9.2-inch. Was it any wonder that the bombardment failed to destroy more than a handful of the many deep dugouts that the Germans had constructed on the Somme? In addition, and unknown to Allied intelligence, the German dugout systems were, in the main, well-designed, well-built and often very extensive networks of rooms and passages, protected by their depth and by the concrete, steel and timber used in their construction.

By the morning of 28 June, with just one day to go, it had been raining on and off for two days. Fearing that the soggy ground would make it difficult for the infantry to advance (the softness of the ground also limited the destructive effects of high-explosive shells, the decision was taken to postpone the attack for two further days. It was thought that the delay would give the guns more time to cut the wire. Instead, the rate of fire was slackened so that the stockpile of shells would last.

At exactly 7.30am on 1 July, having bombarded the German front line for 65 minutes, the guns fell silent as artillerymen adjusted their sights. But before they could open up on the second line, whistles blew in the British trenches and 66,000 heavily loaded infantrymen began to clamber up the ladders and advance through the gaps that had been cut in their wire into no man's land.

Already, German lookouts had alerted their comrades, shaken but safe in their dugouts. Racing out to man their rifles and machine-guns, they at once opened up on the densely packed targets. Some men were hit before they had even made it out of the trenches, others were caught as they bunched in the narrow alleys cut in the British wire. But many more managed to form up into waves and move forward, their rifles at the port, bayonets twinkling in the morning sun. One company of the 8th East Surrey Regiment, attacking near Montauban, even kicked footballs. Its commander, Captain W.P. Nevill, had bought one for each of his four platoons and offered a prize for the first to kick it up to the German trench. It was never claimed; Nevill, like many of his men, was killed short of the enemy's wire.

Whole waves were scythed like hay by the traversing machine-guns. 'Down they all went,' recalled a private of the 11th Suffolk Regiment. 'I could see them dropping one after another as the gun swept along them. The officer went down at exactly the same time as the man behind him. Another minute or so and another wave came forward. Jerry was ready this time and this lot did not get so far as the others.'

Then the German artillery, which had been firing on the British guns and trenches, zeroed in on no man's land. 'Men were falling right and left of me,' remembered a corporal in the 7th Bedfordshire Regiment, 'screaming above the noise of the shell fire and machine-guns – guns we had been assured would have been silenced by our barrage. No man in his right mind would have done what we were doing.'

Those who did make it across no man's land still had to negotiate the German wire. A lucky few, like the 10th West Yorkshire Regiment attacking Fricourt, found it destroyed. Others discovered it had been partially cut and were able to find a way through. But most were confronted with uncut wire defended by well-armed Germans. As they struggled to find a way through, they got snared and were picked off by riflemen at will. 'I could see that our leading waves had got caught by their kilts,' noted a private in the Seaforth Highlanders. 'They were killed hanging on the wire, riddled with bullets, like crows shot on a dyke.'

Perhaps the biggest tragedy occurred in the sector of the New Army's 34th Division. Charged with taking the ruined village of La Boisselle, the divisional commander inexplicably ordered all three of his brigades to advance at the same time. This meant one – 103rd (Tyneside Irish) Brigade – starting from the support line a mile to the rear. Wave after wave of men were machine-gunned as they tried to descend the slope leading to their own front line. A few made it, and even pressed on to the German forward trench and beyond. But the brigade had been destroyed. One battalion had lost 600 men, another 500 (the nominal strength of an infantry battalion was 1,000, 10 per cent of which were kept back out of the battle). The brigadier and two battalion commanders were wounded, a third killed. 'Militarily, the advance had achieved nothing,' wrote John Keegan, the leading military historian. 'Most of the bodies lay on territory British before the battle had begun.'

Not all was doom and gloom. The four divisions making up the right wing of Rawlinson's army, attacking next to the French, had all taken the German front-line trenches. So too had the leading battalions of the 36th (Ulster) Division at Thiepval. Similar gains were made by the Territorial 56th (1/1st London) Division, part of the Third Army's diversionary attack at Gommecourt on the extreme left of the front. However, the neighbouring 46th (North Midland) Division, also a Territorial formation, came up against stiffer resistance and the few small groups that did make it into the German trenches were soon mopped up.

Overall the results of the first hour of fighting were bitterly disappointing. While about a third of the eighty-four battalions that had gone over the top had achieved all their objectives, another third had gained only small and tenuous footholds in the German trenches. The final third had been totally repulsed. Not one of the five villages due to be taken in the first hour had been. Of the 66,000 men who had already gone over the top, half were casualties.

But news reached the rear so slowly that there was little hope of the follow-up attacks by fresh battalions being cancelled. These were mostly piecemeal, their strength and timing depending upon the importance of the sector and the distance to the final objective. As a result, they drew fire from neighbouring sectors and most faltered before they got anywhere near the German trenches. It was now that the 1st Royal Newfoundland Regiment, the only Dominion battalion (although self-governing Newfoundland was not then a province of Canada) to take part in the attack, lost 684 of its 752 men (the highest proportion of casualties suffered that day) in its assault on Beaumont-Hamel to the north of the River Ancre.

By noon the British had committed nearly 100,000 infantrymen from 129 battalions, but their gains were negligible and their losses enormous. Out of thirteen attacking divisions, only one – the 30th – had achieved all its objectives. Advancing to the left of the French, it had captured the ruined village of Montauban. Further north, near Thiepval, the Ulstermen of the 36th Division had taken the formidable Schwaben Redoubt but had been repulsed from the second line. Elsewhere the story was one of tragic failure, with many of the earlier gains wiped out by German counter-attacks. Already, British casualties were in excess of 50,000.

During the afternoon a number of fresh attacks were made, but to little effect. Two exceptions were the 7th Division's successful assault on Mametz – only the second village taken that day – and the neighbouring 18th (Eastern) Division's accomplishment of all its objectives in the sector to the right of Mametz (during which action Captain A.E. Percival of the 7th

Bedfordshires, later famous as the general who lost Singapore, won the Military Cross – see p.63). These gains extended the earlier success of the British right wing to a front of 3 miles. As the French had also taken all their morning objectives on a 3-mile sector between Montauban and the River Somme, the Allies were now in possession of 6 miles of the German front line.

Part of this limited success was due to the fact that in these sectors the artillery 'lifts' were well timed, enabling the infantry to advance with a reasonable amount of protection. They also 'benefited', wrote Keegan, 'from their proximity to the French, whose gunnery, after two years of war, was much superior to the Royal Artillery's, and whose infantry, here belonging to the XX Corps, were among the best soldiers on the Western Front.'

Further north, where the British advanced alone, the preplanned barrage was invariably 'lost' because the infantry arrived in the German front-line trench either too disorganized, too weakened by casualties, or too exhausted to push on at the time required.

The only gain made by either side during the remainder of the day was at Thiepval, where the Ulstermen were finally driven out of the Schwaben Redoubt and back into the old German front line. It would take three months and many more lives before the redoubt was once again in British hands.

By nightfall, the sum total of British achievement was to have taken two fortified villages and moved a 3-mile section of the front line forward by one mile. At no point had the German second-line system of trenches been breached. The cost was a staggering 57,470 casualties (with 19,240 killed and 35,493 wounded); about half the number of men engaged. This was more than the British had lost or would lose in the Crimean, Boer and Korean Wars combined. By contrast, German casualties were just 8,200 (with more than a quarter taken prisoner).

Next morning the attack was continued. But this time only three divisions took part. And so it went on, until the battle finally came to a close in the third week of November 1916 with the British and Dominions having lost some 420,000 men and the French about 200,000, while estimates of German casualties rangedfrom 437,000 to 680,000. There never was a breakthrough. Instead, the front line was simply advanced to a depth of about six miles on a front of roughly twenty miles. Beyond this the German defences were as strong as ever.

'Everything has gone well . . .'

In the early hours of Sunday, 2 July 1916, a messenger arrived at Madame de la Rochefoucauld's house in Amiens with copies of the morning communiqué from Haig's headquarters. Roused, the bleary-eyed occupants rushed to their typewriters. They were British war correspondents, eager to complete the first detailed account of the opening of the Somme offensive.

Accorded the honorary status of officers, and attired as such, the correspondents were particularly conscious of their debt to the Army and their duty to bolster morale at home. They were further hampered by the ubiquitous presence of army censors who checked everything they wrote. Consequently, their reports tended to take the official communiqués at face value – and 2 July was no different.

'Everything has gone well,' wrote one correspondent for a national newspaper. 'Our troops have successfully carried out their missions. All counter-attacks have been repulsed and large numbers of prisoners have been taken . . . Thanks to the very complete and effective artillery preparation, thanks also to the dash of our infantry, our losses have been very slight . . . The first impression of the opening of our offensive is that our leaders in the field have amply profited by the experience of the last two years and that they are directing a methodical and well planned advance.'

A more misleading account of the first day of the Somme could hardly have been written. But it was typical. The Monday (3 July) edition of the *Daily Express*, for example, claimed the capture of 9,500 prisoners (four times the actual total) and many villages, including Contalmaison and Serre (neither of which had been taken). The only report that retained a sense of perspective was the one that appeared in *The Times* – and that was because it was based on the German official communiqué which had been published in a neutral paper.

In truth, the Battle of the Somme was lost before it had begun. To achieve a decisive victory against troops manning a strong defensive position was asking a lot of any army during the First World War. What made it a hopeless task for the British at the Somme was the fact that the battle plan was over-dependent on the effectiveness of artillery fire. When the preliminary bombardment failed to destroy the enemy's wire, much less their deep dugouts, the Germans had plenty of time to man their forward defences before the British could get anywhere near. The attackers' one chance of success lay in surprise, yet they advanced in broad daylight, weighed down with equipment, at a speed of less than 2 miles per hour. More often than not they were left unprotected by the inflexible artillery barrage which kept lifting before they were in a position to follow.

'Strategically,' wrote A.J.P. Taylor, 'the battle of the Somme was an unredeemed defeat . . . The enthusiastic volunteers were enthusiastic no longer. They had lost faith in their cause, their leaders, in everything except loyalty to their fighting comrades. The war ceased to have a purpose. It went on for its own sake, as a contest of endurance.'

The Dieppe Raid

The Dieppe Raid of 19 August 1942 was the largest and most expensive operation of its kind during the Second World War. Of the 5,000 or so Canadian and British troops involved, more than 70 per cent were killed, wounded or captured. They had been sent on a 'mission impossible'.

The initial incentive for such an operation had been to relieve pressure on Russia. In early September 1941, with the German invasion of Russia just six weeks old, Lord Beaverbrook, the influential Minister for Aircraft Production, had urged Churchill either to open a second front on the Cherbourg peninsula or to undertake a super-raid to give Stalin proof of British good will. Within a month, Churchill had appointed Captain Lord Louis Mountbatten, a dashing but incompetent naval officer and distant cousin of the King, as Adviser on Combined Operations.

Mountbatten's task was twofold: to harass the enemy by continuing the programme of raids begun by his predecessor, Admiral Sir Roger Keyes, and to prepare for the eventual invasion of Europe. The raiding programme, in particular, would convince the sceptical Americans of Britain's fighting resolve. It did not seem to matter that the forty-one-year-old Mountbatten – whose only claim to fame was the gallantry he had shown during the quite unnecessary torpedoing of two of his destroyers and the sinking of another – was totally unqualified for the job.

'Neither Beaverbrook nor Churchill seems to have realized that Mountbatten could do real harm and cost lives in a position for which he was utterly untrained,' wrote a historian of the Dieppe Raid, Brian Loring Villa.

In January 1942, Mountbatten's staff drew up a schedule of operations for the year. It included raids on Ostend and St Nazaire in March, Bayonne in April, German-occupied Alderney in May, Dieppe in June and July, and an ambitious attempt on the German headquarters in Paris in August. 'Dieppe was chosen,' according to Captain John Hughes-Hallett, the principal naval planner, 'for no particular reason originally except that it was a small seaport and we thought it would be interesting to do – to capture . . . a small seaport for a time and then withdraw . . . It was not thought to be of any particular military importance . . . And it appeared . . . that it would be about the scale of objective that would be suitable for a divisional attack.'

The stakes were upped in February 1942 when Stalin publically declared that he was willing to negotiate with Hitler's Germany. With much of European Russia already overrun, Stalin was trying to frighten his Western allies into relieving some of the pressure. Hitler certainly thought so, noting in a March directive that the Western Powers would be impelled by 'obligations to allies, and political considerations' to launch raids on mainland Europe. Consequently, he ordered all troops in coastal areas to be kept on high alert.

The Führer's instinct was spot on. In a panic, Churchill asked his senior military advisers, including Mountbatten, to make proposals for a 'sacrifice' operation in France in July to keep Russia in the war. While most suggested temporary bridgeheads or super-raids, Mountbatten resurrected Beaverbrook's idea of seizing the Cherbourg peninsula and holding it indefinitely. Impressed by Mountbatten's audacity, Churchill made him a permanent member of the Chiefs of Staff Committee with the rank of vice-admiral.

As it happened, the Cherbourg expedition was ultimately

rejected as impractical, while most of the raids were cancelled for one reason or another. This still left Churchill needing a propaganda coup to satisfy the Russians, and Mountbatten – with just one small-scale success at St Nazaire in March to his name – needing to justify his appointment. To meet both requirements, Mountbatten resurrected the most ambitious of the raids: a large-scale assault on Dieppe.

It was an unfortunate choice. Just 70 miles across the Channel from Newhaven, the town lies nestled in a narrow gap between high chalk cliffs. The entrance to the harbour is on the extreme eastern end of the seafront; the rest of the town is fronted by a pebble beach that separates the broad esplanade from the sea. To defend this beach the Germans had fortified the two-storey casino at the western end of the esplanade, placed machine-gun nests all along the housefronts, and built pillboxes at both ends. In conjunction with guns sited in caves on both headlands, these strongpoints provided enfilade fire right along the beach. Further protection was provided by a cordon of field batteries some two miles inland, and two heavy six-gun batteries on cliffs either side of the port.

Stationed in and around Dieppe was the second-rate German 302nd Infantry Division, with its headquarters half an hour's drive to the south. In the town itself were two battalions – 1,500 men – of the 571st Regiment, while its other battalions guarded the batteries on the cliffs. Behind the town were stationed cyclist reserves, and within a day's march was the crack SS Adolf Hitler Brigade. Further inland at Amiens was the veteran 10th Panzer Division, recently returned from the Eastern Front.

But while success was unlikely, Mountbatten was more worried about the consequences of inactivity. He could always blame a failed raid on the government or the troops involved. If, on the other hand, the whole summer went by without one significant mission, there would be no one to blame but him. 'He was forced to choose,' wrote Loring Villa, 'between cancelling one again – and thus looking ridiculous – or carrying out a raid that, because thousands of men had been briefed and released to the pubs, might already have been compromised.' He selfishly opted for the latter.

By now the plan to assault Dieppe had undergone a number of changes. At first, Combined Operations had favoured flanking attacks with paratroopers knocking out the coastal batteries and infantrymen advancing from nearby beaches to link up inland. Later it was felt that these attacks alone would not be strong enough to take the town, and so a major frontal assault was added. But this created its own problems: so many simultaneous assaults were bound to cause a naval traffic problem. The agreed solution was for the flanking attacks on the clifftop batteries to precede the main assault by half an hour. As a result, the frontal attack lost the element of surprise.

A further blow was dealt to the expedition in early June when the preliminary bombardment was cancelled. Lieutenant-General Bernard Montgomery – from whose Southern Command the troops would be drawn – and the force commanders had made this recommendation on the ludicrous grounds that it would restore the element of surprise and facilitate the movement of the tanks which were to be landed.

Originally scheduled for June, the raid was postponed after two unsuccessful dry-runs on the Dorset coast emphasized the difficulty of synchronizing so many separate assaults. Bad weather then prevented it going ahead in early July. This time, however, the men had been fully briefed and actually loaded aboard the transport vessels. Fearing a security risk, Montgomery recommended that the raid be cancelled for good. So too did the naval force commander. Both were hastily cut out of the chain of responsibility.

An attempt was made by Mountbatten's staff to conceal intent by launching the invasion fleet from a number of south coast ports. But this simply increased the danger of navigating the mine-free channel, added to the congestion at sea, and made the synchronization of landings even more difficult. The sea-traffic problem had already been exacerbated by the replacement of airborne troops with seaborne Commandos for the task of knocking out the flanking batteries.

The choice of formation to carry out the main attack – the Canadian 2nd Division – was also unfortunate. Comprised mostly of volunteers, it had been in Britain since December 1940 and had yet to see action. This long period of boredom had resulted in low morale and poor discipline, with more than 70 officers and 3,200 other ranks court-martialled for various offences (mainly absence without leave, misconduct and drunkenness). After one particularly violent fight in a pub, the traitor William Joyce, popularly known as 'Lord Haw-Haw', advised his British listeners in a broadcast from Berlin: 'If you really want to take Berlin, give each Canadian soldier a motorcycle and a bottle of whisky; then declare Berlin out of bounds and the Canadians will be there in forty-eight hours.'

On 31 July, Combined Operations issued the preliminary order for the raid – codenamed Operation Jubilee – although it was never formally authorized by the Chiefs of Staff Committee. The final plan was as follows: 3 and 4 Commandos would land at 4.50am on beaches near Berneval and Varengeville, three miles either side of Dieppe, and deal with the two clifftop batteries; simultaneously, four infantry battalions would put in flank attacks from beaches at Puys and Pourville, one mile east and two miles west of the town respectively; half an hour later the rest of the force – three infantry battalions, a tank regiment and A (Royal Marine) Commando – would assault the town head on. Having pushed the tanks as far as the airfield at Arques, about four miles south of the town, the whole force would withdraw.

The great armada of vessels, carrying nearly 5,000 Canadians and 800 Commandos, set sail during the evening of 18 August. Next morning at 1.27am, and again at 2.44am, Hughes-Hallett, the naval commander, received Enigma decrypts warning that an armed German convoy was steaming towards Dieppe. He decided not to cancel the mission. The fact that he was a member of Mountbatten's staff, and one of the architects of the plan, may have had something to do with this decision.

The men of 3 Commando suffered the consequences. At 4am, not far from the shore, their twenty-three Eureka landing craft and gunboat escort had the misfortune to bump into the German convoy. 'The passage through the minefield, which had been cleared by the Navy, was uneventful until nearing the French coast, when a star-shell went up and we were bathed in glorious light,' wrote 3 Commando's signaller.

> My number two, signalman H. H. Lewis, MM, and I were having a pee over the side when this happened. Then all hell let loose. We had been caught by armed trawlers . . . who were escorting a convoy into Dieppe . . . [Our] flotilla was now dispersed over a wide area, smoke and the smell of cordite was everywhere, dawn was breaking and a heavy curtain of fire was coming from the shore where the enemy had been alerted. We were sitting ducks.

Only seven craft survived the unequal firefight and pressed on. Six of them managed to land about 100 men on Yellow Beach 1 near Berneval, but the Germans had been alerted and the commandos failed to get anywhere near the battery. Only one of them made it home; the rest were killed, wounded and taken prisoner or captured unwounded. The other craft, carrying Major (later Brigadier) Peter Young and eighteen men, landed without opposition on Yellow Beach 2 near Belleville. Having scaled the cliff, they engaged the troops defending the battery and kept the guns silent for two hours. Only a lack of ammunition caused them to break off the fight and return to their landing craft. There were no casualties.

A little further west, at Puys, the assault by the Royal Regiment of Canada and the Black Watch of Canada was even more calamitous. Landing fifteen minutes late, and eight minutes after the Germans had sounded the alert, they were cut to pieces on the beach by machine-gun fire from just sixty defenders. So heavy was the fire that many landing craft turned back still laden with troops. Of the six hundred or so men that were put ashore, only six returned home unwounded.

The other flank attack, at Pourville, was marginally more successful. With the Germans only aware of their presence at the last moment, the men of the South Saskatchewan Regiment got ashore relatively unscathed. They came under heavy fire, however, as they attempted to cross the Scie River, where their commanding officer won the Victoria Cross for his gallant leadership. Once over, they continued their advance for a further 2,000 yards until heavy opposition forced them back. By now they had linked up with the follow-up battalion, the Queen's Own Cameron Highlanders, and together they fought on until their ammunition ran out. In all, the Saskatchewans lost 517 men, the Cameron Highlanders 346.

Further west, near Varengeville, 4 Commando was achieving the only tangible success of the day. Half the force under Major Derek Mills-Robert had landed on time at a beach below the Pointe d'Ailly lighthouse with the task of climbing a narrow gully and advancing to the edge of the town, where it would provide covering fire for the assault group led by Lieutenant-Colonel Lord Lovat, which was to silence the battery there. But no sooner had the first force landed than the battery opened fire on offshore shipping. Aware that Lovat's party would not be in position for some time, Mills-Robert decided to attack at once with his own force.

By the time the second group arrived to begin the assault, Mills-Robert's men had been engaging the battery and its defenders for more than half an hour with rifle, Bren-gun and mortar fire. Two officers were killed and many men wounded in the first rush, but the attack was continued by Captain Pat Porteous who led a bayonet charge through the wire to the final position, before being shot in the thighs and falling into one of the gunpits. For this gallant feat he was awarded the Victoria Cross. Having taken and destroyed the battery at a cost of twelve men killed, twenty wounded and thirteen missing, 4 Commando returned to the beaches and re-embarked.

The biggest disaster of the day was the frontal attack on Dieppe. It started well enough with two Canadian battalions, the Royal Hamilton Light Infantry and the Essex Scottish, arriving on schedule, and the Germans momentarily stunned by an air raid. But the fire support from sea and air was hopelessly inadequate and the Germans quickly recovered. As the first landing craft disgorged their men, the defenders opened up with machine-guns, mortars and artillery. Decimated by the withering fire, the survivors of both battalions were forced to take refuge behind the low seawall that separated the pebblebeach from the esplanade.

It did not help that the twenty-seven Churchill tanks of the Calgary Regiment, the 2nd Division's main source of firepower, arrived fifteen minutes late. Even then, with their tracks unable to grip the loose shingle, they presented an easy target and most did not make it off the beach. All eleven that did get over the sea wall and on to the esplanade were knocked out by concealed guns covering the exits from the beach. A few isolated infantry units actually made it into the town, but they could accomplish nothing and were eventually killed or captured.

In an attempt to bolster the assault on the eastern half of the beach, General 'Ham' Roberts, the divisional commander, sent in the Fusiliers de Mont-Royal. Unfortunately, they landed in the wrong place and were quickly pinned down with the rest of the troops which had landed earlier. By now the beach was a deathtrap, swept from end to end by machine-gun fire, bursting shells and flying shingle. The dead and dying lay in heaps.

But still the force commanders would not admit defeat. At 8.30 am, in a desperate last throw, Roberts ordered A (Royal Marine) Commando to land. As they emerged from a smoke-screen, 200 yards offshore, they were met by a murderous fire. Their CO, Lieutenant-Colonel Picton-Phillips, made a brave decision to abort the mission but was killed as he stood up to wave the landing craft away. Saved from total destruction, A Commando (later renamed 40 Commando Royal Marines) still suffered severe casualties.

Half an hour later a general evacuation was ordered. But it was all but impossible under such a heavy fire, and the majority were forced to surrender. At 12.50 pm, when a lone destroyer made a last brave attempt to discover if there were any survivors, all that could be seen were lifeless bodies, burnt-out tanks and landing craft littering the shore.

The final casualty list was appalling. Of the 4,000 Canadians who landed, 800 were killed and 1,900 captured (including 600 wounded). A further 466 naval and Commando personnel were lost, not to mention one destroyer, thirty-three landing craft and all twenty-seven tanks. In addition, 105 RAF fighters were shot down in what proved to be one of the greatest air battles of the war. German casualties were a mere 600 men, including the pilots of 91 planes the British claimed to have shot down (the enemy admitted to 48).

Predictably, Mountbatten put the disaster down to bad luck – bumping into the armed convoy – and to the 'incompetence' of General Roberts 'who wouldn't follow my plan and insisted on a frontal assault without preliminary bombing'. In fact, Mountbatten's staff had approved all the changes to the plan and he himself, for reasons of personal gain, had ensured that the mission went ahead in its final form without the formal authorization of the Chiefs of Staff. One of Montgomery's biographers described Mountbatten as 'a master of intrigue, jealousy and ineptitude. Like a spoilt child he toyed with men's lives with an indifference to casualties that can only be explained by his insatiable, even pathological, ambition'.

Certainly, Lord Beaverbrook never forgave Mountbatten for Dieppe, accusing him at a dinner party of having 'murdered thousands of my countrymen'. Lord Lovat, too, was in no doubt as to where the responsibility for the disaster lay. 'Only a foolhardy commander launches a frontal attack with untried troops, unsupported, in daylight against veterans (who had never known defeat), dug-in and prepared behind concrete, wired and mined approaches,' he wrote. 'It was a bad plan, it had no chance of success.'

Arnhem

Operation Market-Garden, the attempt by Allied airborne troops to seize a corridor through Holland, across the Rhine and into the heart of Germany in September 1944, was always a high-risk venture. But it became positively suicidal once it was known that German panzer formations had moved into the vicinity of Arnhem, where the British 1st Airborne Division had the task of securing the great road bridge over the Lower Rhine.

The sequence of blunders that preceded the doomed operation began on 5 September 1944, when, having captured Antwerp after a lightning dash of 250 miles in five days, Major General Philip Roberts's British 11th Armoured Division stopped to 'refit, refuel and rest'. If he had gone just 18 miles further north, to the narrow neck of the South Beveland peninsula, he would have intercepted the German Fifteenth Army during its flight from the north Belgian coast. As it was, the majority of General von Zangen's army escaped, including two weak SS panzer divisions – the 9th and 10th – which by a freak of fate were sent to regroup in the Arnhem area.

Part of the reason for this omission was that Field Marshal Montgomery (he was promoted from general after D-Day), commanding the British Twenty-First Army Group, had other priorities. The previous day, elated by the successful advance, he had cabled US General Dwight D. Eisenhower, the Supreme Allied Commander for the theatre: 'We have now reached a stage where a really powerful and full-blooded thrust towards Berlin is likely to get there and thus end the war.'

Eisenhower disagreed. Replying on the 5th, he told Montgomery that his aim was to cross the Rhine 'on a wide front' and place a stranglehold on Germany by seizing the main industrial areas of the Saar and the Ruhr. In any case, he said, it was necessary to open the ports of Le Havre and Antwerp, vital for resupplying the Allied armies, before any 'powerful thrust' into Germany could be launched. At the moment, 'no relocations of our present resources would be adequate to sustain a thrust to Berlin . . .'

Montgomery had other ideas, however. Since the formation of the First Allied Airborne Army, six weeks earlier, Eisenhower had been pressing his army commanders to propose plans for a mass parachute drop behind enemy lines. A number of these proposals had been accepted but then cancelled. Montgomery now came up with the idea of capturing the Rhine crossing at Arnhem in northern Holland, more than 75 miles beyond the front line. It was doubly attractive because the Germans had just begun their V2 rocket offensive on London from sites in western Holland.

Eisenhower was briefed by Montgomery on the proposed operation – code-named Market-Garden – at the latter's Laeken headquarters on 10 September. The Germans, explained Montgomery, were expecting him to take the shortest route over the Rhine and into the Ruhr; he would, therefore, surprise them by using a northern 'back-door' route through Holland. Three and a half airborne divisions – the US 82nd and 101st and the British 1st Airborne Divisions and 1 Polish Independent Parachute Brigade – would seize a succession of river crossings and open a corridor for the tanks, followed by the infantry, of the British Second Army. Once across the Rhine at Arnhem, Montgomery's troops could wheel east, outflank the fortified Siegfried Line and penetrate the Ruhr.

Impressed, Eisenhower gave his approval and said that the operation should take place at the earliest possible moment. But he also stressed that the attack was a 'limited one' and 'merely an extension of the northern advance to the Rhine and the Ruhr'.

With the conference over, Montgomery spoke to Lieutenant-General F.A.M. 'Boy' Browning, Deputy Chief of the First Airborne Army, and also GOC of the British Airborne Corps, from which the three divisions involved were drawn. He told him that his paratroopers and glider forces would have to secure, intact, five major bridges – including (from south to north) crossings over the Rivers Maas, Waal and Lower Rhine – over a distance of 64 miles between the Dutch border and Arnhem.

'How long will it take the armour to reach us?' asked Browning, pointing to the most northerly bridge at Arnhem.

'Two days,' replied Montgomery.

'We can hold it for four,' said Browning, adding: 'But sir, I think we might be going a bridge too far.'

He was not the only senior officer to feel uneasy. General Sir Miles Dempsey, commander of the Second Army, had received a number of Dutch resistance reports indicating a growing German presence between Eindhoven and Antwerp – the planned point of attack. One even stated that 'battered panzer formations have been sent to Holland to refit' in the Market-Garden area. Dempsey sent this news to Browning, but because it was not endorsed by Montgomery, who was sceptical of resistance reports, it was not even included in intelligence summaries.

The headquarters staff of the First Airborne Army were more concerned with the fact that Montgomery had given them just seven days in which to prepare for the attack. Previous airborne operations had been smaller, and even so had taken months to devise.

The final plan was for the 'Screaming Eagles' of the US 101st Airborne Division to capture 15 miles of canal and river crossings between Eindhoven and Veghel, while further north the US 82nd Airborne Division took a 10-mile stretch, from Grave to Nijmegen, that included bridges over the Rivers Maas and Waal. The key objective, the great concrete-and-steel bridge over the 400-yard-wide Lower Rhine at Arnhem, was the responsibility of Major-General R.E. 'Roy' Urquhart's 1st Airborne Division – the 'Red Devils' – with Major-General Stanislaw Sosabowski's 1 Polish Parachute Brigade in support. (The railway bridge at Arnhem, as well as a pontoon bridge, were also to be taken if possible.) Once an airstrip had been secured, the 52nd (Lowland) Infantry Division would be flown in to support them. In all, 35,000 airborne troops were involved. But because of a shortage of both powered aircraft and gliders, it would take three days to transport all of them and their equipment. Even then, without tanks or heavy artillery, they would be extremely vulnerable to counter-attack.

A number of things could go wrong: German reinforcements might reach the area sooner than expected; anti-aircraft fire could cause serious casualties; typical autumn weather – fog and high winds – could disrupt the operation; the bridges might be blown; the corridor up which the relieving armour was to race could be cut at any point. If any one element of the precise plan failed, the whole operation failed.

No man was more conscious of this than Browning's intelligence chief, Major Brian Urquhart (no relation to the general). He was 'quite frankly horrified by Market-Garden because its weakness seemed to be the assumption that the Germans would put up no effective resistance', and he had repeatedly voiced his objections to 'anybody who would listen on the staff'. The whole essence of the scheme, as he saw it, 'depended on the unbelievable notion that once the bridges were captured, XXX Corps' tanks could drive up this abominably narrow corridor – which was little more than a causeway, allowing no manoeuvrability – and then walk into Germany, like a bride into a church. I simply did not believe the Germans were going to roll over and surrender.'

Convinced by Dempsey's report that there were panzers in the Arnhem area, he requested low-level reconnaissance sweeps by Spitfires equipped with special cameras. On 15 September, with just two days to go, Urquhart received five photos that clearly showed the presence of German armour. 'It was the straw that broke the camel's back,' he recalled. 'There, in the photos, I could clearly see tanks – if not on the very Arnhem landing and drop zones, then certainly very close to them.'

He rushed to show them to Browning. After carefully studying them, Browning remarked: 'I wouldn't trouble myself about these if I were you.' Then, referring to the tanks, he added: 'They're probably not serviceable at any rate.'

Desperate, Urquhart pointed out that 'whether serviceable or not, [they] were still tanks and they had guns'. He was dismissed and shortly after visited by the medical officer. 'I was told,' he recalled, 'that I was exhausted – who wasn't? – and that perhaps I should take a rest and go on leave. I was out. I had become such a pain around headquarters that on the very eve of the attack I was being removed from the scene.'

That same day, Eisenhower was informed by his Chief of Staff, Lieutenant-General Walter Bedell Smith, that the Dutch resistance had positively identified the 9th and 10th SS Panzer Divisions in the Arnhem area. To Smith's suggestion that the 1st Airborne should be reinforced with another division, Eisenhower lamely replied: 'I cannot tell Monty how to dispose of his troops.' Nor would he 'call off the operation since I have already given Monty the green light'.

He did, however, give his Chief of Staff permission to voice his fears to Montgomery in person. Arriving at Laeken by plane, Smith found the Field Marshal in obdurate mood. 'Monty felt the greatest opposition would come more from terrain difficulties than from the Germans,' recalled Smith. 'He was not worried about German armour. He thought Market-Garden would go all right as set.'

His overconfidence is partly explained by the U-turn performed by the Second Army's intelligence staff. On 16 December, six days after having noted the presence in Holland of 'battered panzer formations', they described the Germans in the Market-Garden area as 'weak, demoralized and likely to collapse entirely if confronted with a large airborne attack.'

Colonel Tony Tasker, the First Airborne Army's chief of intelligence, agreed. Despite the warning from SHAEF (Supreme Headquarters, Allied Expeditionary Force) that two panzer divisions had 'been reported withdrawing to the

Arnhem area', he concluded that there was no direct evidence that the vicinity contained 'much more than the considerable flak defences already known to exist'.

The panzer sightings were not the only intelligence to be ignored. Other elements of van Zangen's army had been reported moving into the area through which the vanguard of the British Second Army – Lieutenant-General (Brian) Horrocks's XXX Corps – was due to advance. While noting the increase in numbers, intelligence officers dismissed the newly arrived German units as 'in no fit state to resist any determined advance'.

Meanwhile, General Urquhart, the 42-year-old commander of the 1st Airborne Division, was having some difficulty choosing suitable landing sites. Ideally, his men would have been dropped by parachute or landed by glider near the Arnhem bridge on both sides of the river. But one bank, the northern, was built up and the other, according to reports, too marshy for men or gliders. Furthermore, thanks to reports from bomber crews of a 30 per cent increase in anti-aircraft fire near the Arnhem crossing, RAF pilots had objected to landing zones close to the bridge. Considering all these factors, Urquhart opted for five sites on heathland and pasture to the west and north-west of Arnhem. These, however, were 6 to 8 miles from the road bridge, a disadvantage exacerbated by the fact that only a part of his division would be landed on D-day, 17 September. 'To seize the main bridge on the first day,' he wrote, 'my strength was reduced to just one parachute brigade.'

His final plan was for two brigades to land on D-day. One, Brigadier P.H.W. 'Pip' Hicks's 1 Airlanding Brigade (glider-borne), would hold the drop zones while the other, Brigadier Gerald Lathbury's 1 Parachute Brigade, would hurry to Arnhem to secure its road, railway and pontoon bridges. Spearheading this assault would be Major Freddie Gough's motorized Reconnaissance Squadron of 275 men with jeeps and motorcycles. Their task was to hold the road bridge until the rest of the brigade arrived.

Brigadier John ('Shan') Hackett's 4 Parachute Brigade and the remainder of 1 Airlanding Brigade would arrive the next day, followed by Sosabowski's 1 Polish Parachute Brigade the day after. Anticipating that by this time the bridge (or even bridges) would have been captured and the flak batteries neutralized, Urquhart had arranged for the Poles to be dropped near the village of Elden, on the other bank about a mile south of the Arnhem road crossing.

Overall, he felt he had 'a reasonable operation and a good plan'. Sosabowski, a 52-year-old former professor at the Polish War Academy, was not so confident. On first hearing about the plan to take Arnhem, he had told Browning that 'this mission cannot possibly succeed'. Asked why, he said that 'it would be suicide to attempt it' with the forces they had. Browning replied: 'But, my dear Sosabowski, the Red Devils and the gallant Poles can do anything!'

When Urquhart – a distinguished infantry officer with no airborne experience – informed his senior officers on 12 September that the division would be dropped 'at least six miles from the objective', Sosabowski was appalled. To reach the bridge most of the troops would have 'a five-hour march – so how could surprise be achieved? Any fool of a German would immediately know our plans.'

He was also dissatisfied with the news that his brigade's heavy equipment and ammunition would go in early by glider to a northern landing zone, while his troops were to land on the opposite bank. What would happen if the bridge had not been secured by then? But he kept these reservations to himself. 'I remember Urquhart asking for questions and nobody raised any,' he wrote. 'Everybody sat nonchalantly, legs crossed, looking bored. I wanted to say something about

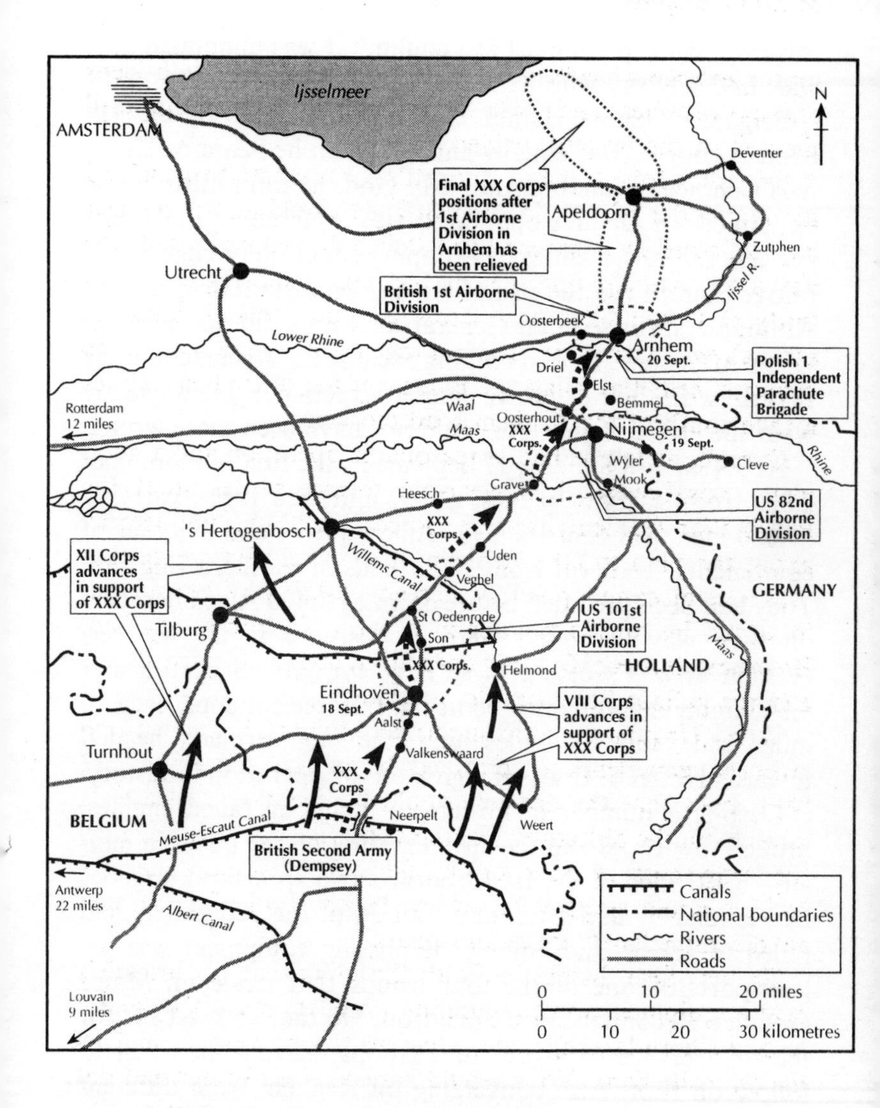

Ijsselmeer
AMSTERDAM
N
Deventer
Final XXX Corps positions after 1st Airborne Division in Arnhem has been relieved
Apeldoorn
Zutphen
Utrecht
Ijssel R.
British 1st Airborne Division
Oosterbeek
Lower Rhine
Arnhem
20 Sept.
Driel
Elst
Polish 1 Independent Parachute Brigade
Bemmel
Rotterdam 12 miles
Waal
Oosterhout
Maas
XXX Corps.
Nijmegen
19 Sept.
Wyler
Cleve
Rhine
Mook
Grave
Heesch
XXX Corps
US 82nd Airborne Division
's Hertogenbosch
Uden
XII Corps advances in support of XXX Corps
Willems Canal
Veghel
GERMANY
Tilburg
St Oedenrode
US 101st Airborne Division
Son
XXX Corps.
HOLLAND
Maas
Helmond
Eindhoven
18 Sept.
Aalst
VIII Corps advances in support of XXX Corps
Turnhout
Valkenswaard
XXX Corps
BELGIUM
Meuse-Escaut Canal
Neerpelt
Weert
British Second Army (Dempsey)
Antwerp 22 miles
Canals
National boundaries
Rivers
Roads
Albert Canal
Louvain 9 miles
0 10 20 miles
0 10 20 30 kilometres

this impossible plan, but I just couldn't. I was unpopular as it was, and anyway who would have listened?'

Brigadier-General James Gavin, commander of the US 82nd Airborne, was equally amazed when he heard of Urquhart's choice of landing sites. 'My God, he can't mean it,' he exclaimed. In Gavin's opinion, it was preferable to take 'ten per cent initial casualties by dropping either on or close to the bridge than to run the risk of landing on distant drop zones'. Although 'surprised that General Browning did not question Urquhart's plan', he said nothing because he 'assumed that the British, with their extensive combat experience, knew exactly what they were doing.' He could not have been more wrong.

At 9.45am on Sunday, 17 September, the first of more than 2,000 troop-carrying planes, gliders and their tugs – carrying 20,000 men, 511 vehicles, 330 artillery pieces and 590 tons of equipment – took off from 24 airfields in southern England. They had been preceded before dawn by 1,400 Allied bombers on their way to pound German positions in the Market-Garden area. Escorting these huge formations were nearly 1,500 fighters and fighter-bombers. In three columns, each 10 miles wide and 100 miles long, the huge armada headed towards Holland. It was the greatest air operation in history.

Despite a number of mishaps – including failed engines, broken tow-ropes and the disintegration of one glider in mid-air – 5,191 men of the 1st Airborne's two spearhead brigades had landed by mid-afternoon. But 36 of the 320 gliders had failed to arrive, and among the missing equipment was the transport for one of the four troops that made up Major Gough's Reconnaissance Squadron. He therefore set off for the road bridge with just three troops, leaving the other to follow on foot. Also converging on Arnhem from different directions were the three battalions of Lathbury's 1 Parachute Brigade. The 2nd and 3rd Battalions, under Lieutenant-Colonels John Frost and J.A.C. Fitch, had the task of supporting Gough at the bridge. Lieutenant-Colonel David Dobie's 1st Battalion would occupy the high ground north of the city.

At 1.30pm, however, Lieutenant-General Willi Bittrich, commanding II SS Panzer Corps, had received word that airborne troops were assaulting Arnhem and Nijmegen. Assuming, rightly, that their objective was to take the road bridges in both cities, he at once ordered the 9th SS Panzer Division to hold the Arnhem area and destroy the enemy troops to its west – 'quick action is imperative' he told the 9th's commander, 'the taking and securing of the Arnhem bridge is of decisive importance' – and the 10th SS Panzer Division to move towards Nijmegen, 'to take, hold and defend the city's bridges'.

Equally fatal to the 1st Airborne Division's chances of success was the realization that its wireless sets were of poor quality and only worked intermittently. Of no use at all were two high-powered radios – operated by a special team of Americans – which were to have been used to call in fighters for close support. It transpired that during the hasty preparations these sets had not been adjusted to the relevant frequency. With the battle barely begun, Urquhart was bereft of air support and out of touch with both Browning's corps headquarters, which had dropped close to Nijmegen, and his forward troops.

Meanwhile, no sooner had the lead tanks of XXX Corps begun their advance into Holland at 2.35pm than they were ambushed by anti-tank fire and nine put out of action. The road was blocked and the break-out halted before it had even begun. Only after rocket-firing Typhoon fighter-bombers and infantry had cleared the German positions, and an armoured bulldozer had pushed the burning tanks off the road, could the column resume its advance. But German opposition was continuous and, having expected to cover the 13 miles to

Eindhoven 'within two to three hours', XXX Corps had only managed the 7 miles to Valkenswaard by nightfall.

Even more unfortunate was the discovery by the Germans, in the remains of a glider that had crashed near Vught in central Holland, of a briefcase containing the entire Market-Garden operational plan. By evening, Field-Marshal Walter Model, commanding Army Group B, was in possession of the names of the formations involved, the bridge objectives, the location of the landing and drop zones, and the timing of the resupply lifts. Although not entirely convinced that the plan was genuine, Model (who had actually had to flee his HQ on the outskirts of Arnhem when 1st Airborne began to land west of the city) did at least alert all anti-aircraft units about the drops due to take place a few hours later.

Back at Arnhem, Gough and the 1st and 3rd Parachute Battalions had run into SS panzer units before they had gone more than 2 miles. Everything now depended upon Frost's 2nd Battalion, taking the most southerly of the three routes to the road bridge. It, too, was held up for a time but managed to occupy the northern end of the road bridge by 8pm. (The railway bridge was blown by the Germans just as one of Frost's companies arrived to secure it, and the enemy had removed the centre section from the pontoon bridge). Unfortunately, an attempt to seize the southern end was beaten back by a group of SS Panzer Grenadiers that had moved into position only half an hour earlier. That night, two more attempts were made to secure the other side of the bridge. Both were repulsed. By dawn, the arrival of stragglers from the two missing parachute battalions, as well as Brigade Headquarters (though without Brigadier Lathbury, who had become separated from his HQ) and Gough with some of the Reconnaissance Squadron, had swelled Frost's force to between 600 and 700 men. But he was effectively cut off from the rest of the division and German armour was on its way.

The second day of the battle, 18 September, was a catalogue of disaster for the British. Anxious to know what his forward troops were up to, Urquhart had spent the first night with Brigadier Lathbury and the 3rd Parachute Battalion at Oosterbeek on the western outskirts of the town. Next morning, as the battalion continued its advance towards the bridge, it was ambushed by elements of the 9th SS Panzer Division and Lathbury was wounded. Urquhart and two other officers were forced to spend the day and night hiding in an attic. Frost, meanwhile, was desperately fighting off determined SS attacks from the north, east and south. During the latter assault over the bridge by the 9th SS Reconnaissance Battalion, almost all the twenty-two armoured cars and half-tracks involved were destroyed.

Unfortunately, Hackett's 4 Parachute Brigade, due to arrive at 11am, had been delayed by fog and did not land until mid-afternoon, when only one battalion was deployed in the attempt to break through to Frost at the bridge. Part of the problem was a confusion as to who was in command. Urquhart had left orders that if either he or Lathbury was incapacitated, Hicks was to take over. But Hicks, though senior in age, was junior to Hackett as a brigadier, and the latter was not pleased to find Hicks commanding the division. The end result was that the remaining two battalions of Hackett's brigade did not join in the battle for the bridge until the morning of the 19th, by which time Urquhart had rejoined his divisional headquarters (by then established in the Hartenstein Hotel in Oosterbeek).

His first act was to send a signal altering the Polish Parachute Brigade's landing zone and the supply dropping point, both of which were now outside the divisional area. But the message got through too late; most of the supplies were intercepted by the Germans, while the glider-borne element of Sosabowki's brigade was badly cut up when it landed north of

Oosterbeek on the 19th. The Polish parachutists, on the other hand, were delayed by fog and did not take off for another two days.

All might not have been lost if the tanks of XXX Corps had managed to reach Arnhem within the forty-eight hours allotted. But determined German resistance and the blowing of a bridge over the Wilhelmina Canal at Son had slowed their advance. During the morning of 19 September, they arrived at the southern outskirts of Nijmegen only to discover that units of the 10th Panzer Division were in control of the road bridge. That afternoon, a joint assault by the Guards Armoured Division and the 82nd Airborne Division was repulsed with heavy losses.

Next morning, having given up all hope of breaking through to the surviving Arnhem bridge, Urquhart ordered his remaining forces to concentrate around the Hartenstein Hotel in Oosterbeek, not far from the river. By nightfall, they had formed a perimeter 2,000 yards deep and 1,000 yards wide. It was, wrote one of his officers, 'by no means an ideal position for the airborne troops, the thick woods and narrow streets being difficult to defend with the limited number of men left'.

At the bridge, meanwhile, the 2nd Battalion's heroic defence was coming to an end. Frost had been hit in the legs by a mortar bomb and, along with the rest of the wounded, was evacuated that evening during a truce. Gough took over, but the dwindling command was finally overrun the following morning, 21 September. More than half of those who had defended the north end of the bridge had by then been killed or wounded, and many of the survivors were captured. They had held the crossing for three days and four nights.

In an epic assault the previous evening, the 82nd Airborne Division and the tanks of XXX Corps' Guards Armoured Division had taken the road bridge over the Waal at Nijmegen (the railway bridge was also captured intact). But instead of pressing on – with the 10th SS Panzer Division in temporary disarray and the 11-mile stretch of road to Arnhem open – the leading tanks were ordered to halt while infantry support was brought forward. By the time they got under way the following morning, the Germans had retaken the bridge at Arnhem and were rushing reinforcements across it to form a defensive line at Elst, 4 miles to the south.

At 5.15pm on 21 September, the parachutists of Polish Brigade were dropped at a rearranged zone on the south of the Lower Rhine, opposite Urquhart's bridgehead. But the ferry there was no longer operating and they were forced to take up defensive positions in nearby Driel. That night, they made repeated efforts to cross the river and bolster the bridgehead – but only 250 made it across, and many more became casualties.

On Sunday, 24 September, the forward elements of XXX Corps' 43rd Division linked up with Sosabowski at Driel. But by now the remnants of Urquhart's force were out on their feet. Under constant shell and mortar fire, weakened by persistent attacks from infantry, tanks and self-propelled guns, the area they occupied was ever-diminishing. Earlier that day the Germans had agreed to a truce so that the wounded could be taken to hospitals in Arnhem. Others surrendered without orders. For those left, the smell of decomposing corpses, unwashed bodies and makeshift latrines was almost unbearable.

Horrocks, the commander of XXX Corps, still hoped to reinforce the bridgehead by attempting a major crossing further downstream. But General Dempsey, from his headquarters near Eindhoven, overruled him. 'Get them out,' he ordered.

That night, to facilitate the withdrawal, the 4th Dorsetshire Regiment were ordered across the river. Many were lost in the water, even more on the opposite bank and only a handful reached the airborne perimeter.

During the night of the 25th, having left the badly wounded in the care of the doctors and padres, and under the cover of an artillery barrage, Urquhart and the remnants of his force were ferried across the river. Of the 10,005 who had been deployed on the north bank, only 1,741 officers and men of the 1st Airborne Division, 422 glider pilots and 160 members of 1 Polish Parachute Brigade made it to safety. Accompanying them were 75 Dorsets, making 2,398 survivors in all. The total casualties were 1,200 killed and and 6,378 missing, wounded and captured. At the same time, the Germans lost 3,300 men, including 1,100 dead.

Operation Market-Garden was an ambitious venture that was doomed to fail because of a number of blunders committed during the planning stage. Not the least of these was the choice of landing sites so far away from the road bridge at Arnhem, and the assumption that an armoured column could advance 64 miles into enemy-held territory on a single highway in just forty-eight hours. But what made disaster certain was the rejection of more than one source of intelligence that two veteran SS panzer divisions had recently arrived in the Arnhem area.

'It began to seem to me,' wrote one paratrooper during the battle, 'that the generals had got us into something they had no business doing.'

'The division's gone'

As the bedraggled band of survivors from Oosterbeek trudged through the darkness and rain to Driel, Captain Eric Mackay of the 1st Parachute Squadron, Royal Engineers, scanned every face. Having escaped from Arnhem, where he had performed heroics in defence of the schoolhouse near the bridge, he was looking for members of his squadron.

'The worst thing of all was their faces,' he recalled. 'They all looked unbelievably drawn and tired. Here and there you could pick out a veteran – a face with an unmistakable "I don't give a damn" look, as if he could never be beaten.'

He stayed by the road all night, but to no avail. 'I didn't see one face I knew. As I continued to watch, I hated everyone. I hated whoever was responsible for this and I hated the army for its indecision and I thought of the waste of life and of a fine division dumped down the drain. And for what?'

When it was light, Mackay returned to Nijmegen to search collecting points and billets. Of the 200 engineers in his squadron, only 5, including himself, came back.

Padre Pare of the Glider Pilot Regiment had spent the night in the shell-blackened Schoonoord Hotel, one of the main casualty stations in the Oosterbeek perimeter, oblivious to the evacuation. He awoke with a start, aware that something was wrong. It was too quiet. Hurrying out of the room, he noticed a medic standing recklessly at a window. As he approached, the medic turned and said: 'The division's gone.'

'You're mad, man,' replied a disbelieving Pare.

'Look for yourself sir,' said the medic, shaking his head. 'We really are prisoners now. Our chaps have had to retreat.'

As Pare struggled to come to terms with the awful tidings, the medic continued: 'Sir, you'll have to break the news to the patients. I haven't got the nerve to tell them.'

Steeling himself, Pare began to make the rounds.

cont. overleaf

'Everyone tried to take it in good heart,' he recalled, 'but we were all in a fit of depression.' Then a soldier sat down at a piano in the large room that was housing most of the wounded and began to play a selection of popular tunes. The room's occupants began to sing along.

'It was queer after the hell of the last few days,' said Pare. 'The Germans could not understand it but it was easy enough to explain. The suspense, the sense of being left behind produced a tremendous reaction. There was nothing left to do but sing.'

Bravo Two Zero

On 22 January 1991, during the Gulf War, an eight-man SAS patrol – code-named 'Bravo Two Zero' – was inserted deep behind enemy lines in western Iraq. The extraordinary exploits of one corporal have rightly gone down in Special Forces folklore. But overall the operation was a costly flop, condemned to failure from the outset by faulty intelligence, inadequate equipment and human error – deficiencies not normally associated with the 'Regiment'.

The war had begun five months earlier, in the small hours of 2 August 1990, when Iraqi tanks and troops invaded neighbouring Kuwait. Within weeks, the United Nations' Security Council had condemned the aggression and a powerful Coalition force of Western and Arab powers had begun to form in Saudi Arabia. Their mission: to liberate Kuwait and its valuable oilfields.

At first, only two SAS squadrons – A and D – were dispatched to the Middle East to prepare for operations. Then, in early January 1991, shortly before the start of the

Coalition's air offensive against Iraq, they were joined by half of B Squadron. Among the new arrivals was 28-year-old Corporal Chris Ryan, a tough Geordie who had trained with the Territorial SAS as a boy. Before leaving the SAS lines in Hereford, he had asked his troop colour sergeant if he could take with him some cold-weather mountaineering gear.

'Nah,' came the dismissive reply. 'You're going to the fucking desert, yer dick! It won't be cold there.'

Little did the ill-informed colour sergeant realize how harsh the Iraqi winter could be.

Last to arrive, the B Squadron troopers were given the remnants of the equipment, the men from A and D Squadrons having had the pick of vehicles, kit, weapons, communications systems and other specialized matériel. The seriousness of the situation was not realized until build-up training in the Saudi Arabian desert. Ammunition was scarce and 'contact' drills – how to react on bumping into the enemy – in which, normally, hundreds of rounds would be expended, were limited to one per man. 'We seemed to be short of everything,' recalled Ryan, 'not least proper desert vehicles.'

Instead of purpose-built 'Pinkies', long-wheelbase Land-Rovers with mounts for heavy machine-guns and anti-tank missiles, they were issued with 'Dinkies', the short-wheelbase version with no gun mountings or seat belts. After stripping them down, they took them out in the desert to practise navigation and vehicle contact drills. It was a fiasco: the Land-Rovers kept getting stuck in the sand, machine-guns could not be fired with any accuracy, and men were thrown out as drivers reversed hard into an emergency J-turn.

'We laughed at the time,' recalled Ryan, 'but underneath we were alarmed by the thought that we might be sent across the border in these vehicles and might have to use them for real . . . In general, it was pathetic trying to operate with the wrong equipment, and altogether our training was poor.'

It had still not been decided how to deploy B Squadron, although a number of hare-brained schemes were proposed by the planners. They included: parachuting into Kuwait City, capturing a tower block and directing mortar and artillery fire on to Iraqi positions; and jumping into Baghdad itself to blow up key installations. Neither, thankfully, were authorized.

Instead, Lieutenant-General Sir Peter de la Billière, the British commander and a former SAS Commandant, managed to persuade US General Norman Schwarzkopf, the Commander-in-Chief of the Coalition forces, to insert SAS patrols deep inside enemy territory. His plan was to deploy A and D Squadrons as four heavily armed motorized patrols to locate and destroy the mobile Scud missiles that the Iraqis were using against Israel. Before that, however, three eight-man patrols from B Squadron – Bravo One Zero, Bravo Two Zero and Bravo Three Zero – would infiltrate deep into Iraq and set up observation posts to provide valuable intelligence.

Bravo Two Zero's mission was to lie up near a major road known as the Main Supply Route (MSR) in western Iraq for fourteen days. During that time it was expected to sever Iraqi fibre-optic communications and report on the movement of Scuds on their mobile launchers. But in choosing their area of operation – a bend in the MSR, 30 kilometres from an airfield on one side and the town of Banidahir on the other – the patrol had to make do with a tactical air map that only showed major features. Intelligence about the region was virtually non-existent, and there was no way of knowing how close they would be to small civilian settlements and military installations.

The task was further complicated by the lack of adequate equipment. 'Look,' complained one disgruntled trooper to his officer, 'we've got no pistols, we've got no grenades, we've got no claymores [anti-personnel mines]. You're expecting us to make claymores out of ammunition boxes, and it's bloody stupid, because they're not effective.'

'Listen,' came the vexed reply, 'you're about to go to war. You'll take whatever you can get. You're not in Northern Ireland, where you can ask for any kind of asset. You've got to improvise.'

The next question was whether to use vehicles. On the one hand they would give the patrol greater mobility and enable it to drive out of trouble if things went wrong; on the other hand they would be hard to conceal and might give the lying-up position (LUP) away. The other two patrols decided to take the 'Dinkies'. Bravo Two Zero, perhaps mindful of the vehicles' inadequacies, opted to go in on foot. However, this meant that each man had to carry extra equipment; the average load, including bergen (pack), belt-kit, weapons, extra ammunition and a jerry can of water, was a staggering 120 kilos apiece. Sleeping bags were left behind because there was no room and because the weather was forecast as likely to be clement for the duration of the mission.

Finally, on the evening of 22 January, after one false start, the patrol was inserted into Iraq by a low-flying Chinook helicopter and dropped off at a point 2 kilometres south of the MSR. Each man was wearing standard-issue DPM (disruptive pattern material – i.e. camouflage) combat fatigues and lightweight, sand-coloured desert smocks that dated from the Second World War. Half the patrol were armed with 203s (American M16 automatic rifles with grenade launchers mounted beneath); the other half with Minimi 5.56mm light machine-guns. All carried disposable 66mm rocket launchers. For escape purposes, they were equipped with a silk map, twenty gold sovereigns and a letter from the British government – the text in English and Arabic – promising to pay £5,000 to anyone who delivered its bearer safely back to the Coalition forces.

Just minutes after landing, the patrol heard barking and could see a building and a water tower about 1,500 metres to the east. 'It shouldn't have been there, but it was,' recalled Sergeant Andy McNab, the patrol commander, a former soldier of the Royal Green Jackets from Peckham who had joined the SAS on his second attempt in 1985. 'I was worried about how good the rest of our information was. But at the end of the day we were there now. There wasn't a lot we could do about it.'

Taking it in turns to carry the kit and provide security, they reached the bend of the MSR in the early hours of the following morning. After a brief recce, McNab selected a lying-up point under an overhang at the end of a deep wadi (dried-up watercourse – except in the rainy season). But it was only really secure from two sides. If anyone followed their tracks up the wadi, or looked over its lip at a point opposite the overhang, they would be bound to be seen. 'This is no bloody good,' somebody muttered.

At first light there was more bad news. The MSR was 200 metres to the north, but a further 200 metres to the north-west was an Iraqi anti-aircraft position. 'The sight gave us a nasty jolt,' wrote Ryan, 'to see those guns so close to us was bloody frightening, because they could only have been placed there to protect some installation from air-attack, and they showed that we were right on the edge of an enemy position.'

It was now vital to get a message back to base requesting a relocation or a return. But when 'Legs' Lane, the signaller, tried to use the radio there was no response. Throughout the day he made repeated attempts, using different frequencies and a variety of antennas. All to no avail. What the patrol did not realize until much later was that, incredibly, they had been given the wrong frequencies. There was no panic, however, for the lost-communications procedure meant that a helicopter would return to the dropping-off point forty-eight hours later with a new radio or fresh orders.

That night, McNab and three others went on a long recce.

They were horrified to discover, in addition to the AA guns, plantations to the north and south, and civilians further down the road. 'From a tactical point of view,' wrote McNab, 'we might as well have sited our LUP in the middle of Piccadilly Circus.'

The plan now was to rendezvous with the helicopter the following night, but first they had to hide out for another day. A tense morning passed uneventfully. Then, around mid-afternoon, they heard a young boy calling his goats and froze as the clatter of hooves and the tinkling of a bell got closer. First the head of one goat appeared over the edge of the wadi, followed by more. Finally, the boy's head and chest came into view. Fortunately, he was on the same side as the overhang and there was just a chance he would not spot the patrol.

It was Sergeant Vince Phillips, the second-in-command, who gave the game away. Anxious to see what was going on, he eased his head up until the boy caught sight of him. The calling stopped and the only sound was the patter of disappearing footsteps. They had been compromised.

At once the patrol spread out in a defensive position at the end of the wadi as 'Legs' tried to send an emergency message. But before long the ominous sound of tank tracks could be heard advancing up the wadi. 'We were stuck,' recalled Ryan, 'pinned like rats in the dead-end of the ravine.'

As they waited they checked their kit and gulped down water and chocolate, not knowing when they would have another opportunity to eat. Suddenly, the vehicle burst into view – it was not a tank but a bulldozer, with its blade raised in front like a shield. Spotting the patrol from a distance of 150 metres, the driver stopped, reversed out of sight and drove off.

Ideally, the SAS troopers would have waited until dark before leaving their cover. But they had been spotted and Iraqi troops were bound to be on their way. Calling the patrol together, McNab said: 'We're going to move from here. We're going to go west, try to avoid the AA guns, and then head south and go for the RV [rendezvous] with the helicopter.'

Leaving the surplus kit behind, they shouldered their bergens and set off in single file, with Arab shamags around their heads in an attempt at disguise. As the wadi petered out into flat plain, the patrol continued west for five minutes. Still no sign of the enemy. But as they turned south, Ryan, the lead man, noticed two men with rifles on high ground to their left. Pointing them out, he urged the rest of the patrol to keep going.

'Then I blew it in a big way,' recalled Ryan. 'I thought, "I'm going to try the double bluff here," and I waved at them. Unfortunately I did it with my left hand, which to an Arab is the ultimate insult – your left hand being the one you wipe your arse with.'

One immediately brought up his weapon and opened fire. The patrol responded with short bursts before moving off, only stopping to return fire. Before long the two Iraqis were joined by an armoured car equipped with a machine-gun and a truck carrying a dozen troops. As the enemy spread out and opened up, the patrol replied with rockets, one blowing the truck to pieces. Then the patrol was up and running, occasionally swivelling to shoot at the enemy. During one pause, Ryan got out his emergency tactical rescue beacon (TACBE) and tried to contact an American AWACS (airborne warning and control system – a long-range aircraft whose purpose is electronic surveillance of enemy radar, signals, movements etc.) McNab was doing likewise. There should have been an immediate response – but there was nothing.

'My TACBE's fucked!' yelled Ryan.

'Can't get through on mine either,' said NcNab.

They kept going and eventually someone shouted: 'I'm ditching my fucking bergen!' The rest followed suit, a machine-gun bullet thudding into Ryan's pack as he struggled to release it.

By now the Iraqi anti-aircraft guns had joined in the unequal fight. Tracer bullets and shells dogged the eight SAS men as, exhausted and no longer able to run, they struggled uphill. At last they got over the top and into dead ground. 'Fucking hell!' gasped McNab, 'I don't know how we managed that.'

'Nor do I,' Ryan replied, offering him a swig of whisky from the flask he had gone back to retrieve under fire from his abandoned bergen. Incredibly, everyone had made it without so much as a scratch. A group decision was quickly reached. If they went for the helicopter RV, there was a good chance it would be ambushed. They decided, instead, to make for the Syrian border, 120 kilometres to the west. First, however, they would head south to put the Iraqis off their track, before turning west and then north towards the Euphrates, where they would find water. They would then follow the river westwards towards Syria.

As they came out of the depression, more machine-gun and anti-aircraft fire was directed towards them – but it was wildly inaccurate in the failing light. A couple of hours of speed marching later and two members of the patrol were struggling. Vince Phillips had hurt his leg in the contact and was limping. Stan, a Rhodesian and one of the strongest members of the patrol, was suffering from heat exhaustion. He had not had time to change out of his thermal underclothes and, despite the freezing temperature, had sweated himself dry. They stopped for a while and forced him to drink lots of water with rehydrate powder. But still he was tired and disorientated.

'Listen,' said Ryan, 'if you don't start walking we're going to fucking leave you. Understand?'

Stan grunted his assent.

'Get up, then,' continued Ryan. 'Get on my arse and don't

leave it. Just fix your eyes on my webbing, and keep that in sight.'

Having shared out Stan's equipment between them, the patrol moved off. Ryan was leading, Stan next, followed by Vince and then McNab, then the rest. At every halt, Mark, a New Zealander, would take a fix with his hand-held GPS (global positioning satellite) receiver. After 16 quick kilometres south, and 10 west, they turned north. By now the strain was beginning to tell. Moving at speed with 50-pound belt-kits and their weapons, they were all sweating heavily and had soon drunk most of their water.

At a point about seven miles south of the MSR, they stopped for another position check. Ryan suggested pushing on as fast as possible until they were over the MSR and on to the high ground beyond. McNab agreed. But a few minutes after setting off again, McNab heard jets coming from the north. He made an instant decision, putting his hand on Vince's shoulder and saying, 'We're going to stop and try TACBE.'

Vince nodded and said, 'Yep, OK, yep.'

By the time McNab's frozen fingers had located the TACBE, the last couple of jets were flying over. 'Hello any callsign,' he transmitted, 'this is Bravo Two Zero, Bravo Two Zero. We are a ground callsign and we're in the shit. We have a fix for you. Over.'

'Say again, Bravo Two Zero,' an American voice replied. 'You're very weak. Try again.'

'Turn back north,' said McNab, realizing the jet was flying out of range. But it was too late. The pilot had continued on towards his objective and McNab could only hope that he would report the brief conversation.

As he rose to his feet, he realized with horror that the leading three members of the patrol had disappeared. Vince had obviously been too done in to register the order to halt. McNab silently cursed himself for not checking that the others had stopped. He decided to keep on the same compass bearing in the hope of bumping into the missing trio.

Ryan, still tramping on with Stan and Vince, was unaware of the split until he had crossed the MSR. 'Where the fuck's the rest of the patrol?' he asked Vince.

'I don't know,' came the weary reply. 'We've lost them.'

'What do you mean, lost them?'

'They split off somewhere.'

Frantically, Ryan used Stan's nightsight to scan the open gravel plains. There was nothing. Waiting five minutes until midnight, he used his TACBE in the knowledge that anyone in difficulty would listen out every hour and every half-hour. Again nothing. Thirty minutes later, after a second fruitless attempt, he decided to keep going; he had to put as much distance between his group – with just two M16s and a bayonet between them – and the MSR as possible. McNab's group had the satellite navigation equipment, and so Ryan was forced to rely on map, compass and dead reckoning.

The next day was torture. For twelve hours, Ryan and his companions lay hidden in deep ruts made by tank tracks and almost froze to death. It snowed, but there was a small military outpost just 600 metres away and they could do nothing to warm themselves. 'The cold bit into our very bones,' wrote Ryan. 'Time slowed to a crawl.'

At about 5.30pm, as it began to get dark, they were able to stand up and improve their circulation. All three men were suffering from faltering speech and slowness of movement – the early signs of hypothermia. Realizing that they had to increase their blood circulation or they would die, Ryan led them due north. It started snowing again and gradually Vince began to lag behind.

'I want to go to sleep,' he implored. 'I'm too tired.'

'Vince,' said Ryan, 'we can't sit down. If we stop, we're going to fucking DIE. Get that?'

But despite Ryan's cajoling, Vince continued to struggle. Ryan and Stan, feeling a little better now, would forge ahead and then wait for Vince to catch up. Eventually he failed to appear. They retraced their steps for while, but to no avail. 'Stan,' said Ryan finally. 'We've got to leave him, or we'll kill the pair of us.'

'OK, then,' said Stan, 'Fair enough.'

(Vince died from hypothermia that night. After the war his body was returned to the UK for burial.)

Having got down off the high ground north of the MSR, they hid in a shallow wadi. Luckily, the sun came out, which probably saved their lives. Around midday, however, they were discovered by another goatherd, this time a young man in his twenties. Ryan was all for killing him, but Stan would not agree. After trying to communicate by sign language, Stan decided to go off with the Iraqi in the belief that he would lead him to a vehicle.

'Don't do it!' warned Ryan, reminding him that even Iraqi civilians were the enemy.

But Stan had made up his mind. 'It's OK,' he said. 'I'll take the risk and go with him.'

Seeing that further argument was useless, Ryan told him that he would wait until 6.30pm before setting off on a northerly bearing.

After walking for four hours, Stan was led into a trap and overpowered – but only after he had shot three Iraqis. These same troops then set off after Ryan in two four-wheel-drive vehicles. He stopped one with a rocket, the other with a 203 grenade. That night he reached the Euphrates and four days later, after numerous close shaves, he crossed the Syrian frontier.

Having survived on just two small packets of biscuits and virtually no water, he had covered more than 300 kilometres in eight days. It was two weeks before he could walk properly, six weeks before he could feel any sensation in his fingers and toes. He had a blood disorder from drinking dirty river water and had lost 36 pounds. His body had literally been feeding on itself.

'It was,' wrote McNab generously, 'one of the most memorable E&Es [escape and evasions] ever recorded by the Regiment, as far as I am concerned ranking above even the legendary trek through the desert of North Africa by Jack Sillitoe, one of David Stirling's originals, in 1942.' (Stirling founded the SAS.)

McNab, too, had been within a hair's breadth of getting away. The day after the patrol had split, his group had lain up up in the lee of a mound. It was bitterly cold, however, and Mark was showing signs of exposure. They therefore decided to risk a daylight move, but 3 miles beyond a metalled road they turned back. Short of water and close to collapse, they elected to hijack a car and attempt to drive to the border.

A Close Shave

As well as searching for Scuds and observing the main supply routes in Iraq during the Gulf War, the SAS carried out a number of important missions against the Iraqi communications network. One came within a hair's breadth of disaster.

Transported at night by Chinook helicopter, the 16-man SAS team's objective was to destroy a radar dish deep inside Iraqi territory. At first all went well. Satellite and low-level photo reconnaissance had revealed minimal enemy activity and the absence of anti-aircraft fire seemed to bear this out.

Crowded in the hold, reliant on the skill and nerve of pilots flying with night-vision goggles, the anxious

cont. overleaf

troopers were relieved to hear they were approaching the landing zone. But as the rear wheels of the aircraft made contact with the ground, a huge blast sent shrapnel through the floor of the hold, tore away part of the landing gear and punctured the tyres. They had touched down in a minefield.

While the pilot struggled to keep airborne and avert catastrophe, the SAS team prepared to jump off the rear ramp. They had quickly come to the selfless conclusion that the only way to save the heavily laden Chinook was to reduce its payload. But the pilot would have none of it. Telling the SAS men, via a crewman, to stay where they were, he strained every sinew to build up enough power to lift off. The gamble paid off. A few nerve-tingling seconds later, the badly damaged helicopter was airborne and limping back to its Saudi base.

Incredibly, given that no fewer than 77 holes were later counted in the Chinook's fuselage, there were no casualties. But for the heroics of the senior pilot and his crew, it could have been far worse. His gallantry was recognized with the award of the Air Force Cross; his crew had to make do with cold beers courtesy of 16 very grateful SAS men.

With one of them faking injury, they stopped the first vehicle to happen along – a yellow minicab. Its occupants were unceremoniously ejected and the five of them piled in – McNab driving. But with 13 kilometres to go to the border, they were forced to abandon the car at a military checkpoint and shoot their way clear. Now on foot, and with every man and his dog looking for them, they were moving through a vast military installation, just 7 kilometres from safety, when a firefight erupted. The patrol split into three. On his own, Bob Consiglio held off the Iraqis for some minutes until he was hit in the head by a bullet that exited through his stomach and ignited a white phosphorous grenade in his webbing. He died instantly.

'Legs' Lane and 'Dinger' tried to escape by swimming the Euphrates. Incredibly, both made it across the 500-metre stretch of freezing water, but 'Legs' was done in and had to be left. He died of hypothermia, and 'Dinger' was captured soon afterwards.

McNab was the last to be captured. After Mark was shot in the foot, he carried on alone and, having run out of ammunition, was picked up by Iraqi commandos just 2 kilometres short of the border. All four prisoners were brutally tortured before being repatriated at the end of the war.

Bravo Two Zero became one of the most decorated patrols in SAS history. McNab was awarded the Distinguished Conduct Medal; Ryan, Consiglio and Lane the Military Medal. But it was also one of the most costly, with three of its eight men dead and four captured – a casualty rate of 87 per cent. And all because of amateurish blunders – insufficient intelligence and the wrong radio frequencies, in particular – that were made before the operation had even begun.

Chapter 3

Meddling Ministers

'War is nothing but the continuation of politics with the admixture of other means,' wrote Karl von Clausewitz, the great nineteenth-century Prussian military theorist. Therein lies the paradox of war. Having brought it about, politicians are then expected to take a back seat while it is prosecuted by professional soldiers. But many cannot resist interfering, particularly those who subscribe to Talleyrand's belief that 'War is much too serious a thing to be left to the military.'

They meddle for a number of reasons: some are faced with immediate political pressures; others have longer term political aims. But the most dangerous ones are those who see themselves as masters of strategy, and even of battlefield tactics. They tend to intervene in the day-to-day running of a war – and the effect is invariably fatal.

Bannockburn

King Edward II of England belonged to the latter category. During the two-day Battle of Bannockburn (23–24 June 1314) he consistently ignored the advice of his more experienced captains and attempted to defeat King Robert the Bruce's smaller Scottish force with his knights alone. It was a blunder of epic proportions, and his huge army was destroyed as a result.

Born at Caernarfon Castle in April 1284, Edward was the

fourth son of Edward I by his first wife, Eleanor of Castile. He could not, however, have been less like his illustrious father, Nemesis of Sir William Wallace ('Braveheart') and 'Hammer of the Scots'. Though tall, handsome and exceptionally strong, the younger Edward had little interest in the business of war and preferred the more mundane activities of digging trenches and thatching houses. He had a passion for fine clothes, indulged in heavy drinking bouts, and preferred the company of men, particularly those of low origin and vulgar tastes.

The one exception was Piers Gaveston, the son of a Gascon knight who, by his faithful service, had earned the favour of Edward I. Young Edward's childhood playmate, Gaveston had gained an early influence over the shallow prince that was to last into manhood. It was almost certainly cemented by a homosexual relationship.

The King well understood Gaveston's malign influence and banished him in April 1307 for enlisting the prince's support in his quest to be given the French county of Ponthieu. When Edward I died, on his way to put down a rebellion led by the new Scottish monarch, Robert the Bruce, in July of the same year, Gaveston was granted a temporary reprieve. But the powerful English barons were no more reconciled to Gaveston than Edward I had been, knowing that he supplanted them in the young King's councils, and they insisted that the banishment order be upheld.

Edward II refused, and instead heaped more titles, appointments, lands and riches upon his favourite. By 1312 the barons – chief among them the Earls of Lancaster, Warwick and Arundel – had had enough. Securing Gaveston under the promise of a safe conduct, they summarily executed him. The king was outraged and, joining forces with his loyal barons, marched against Lancaster and his fellow conspirators.

Robert the Bruce was not slow to take advantage of the riff in the English leadership. At war with England, on and off, since his coronation in March 1306, he convened a parliament at Ayr in July 1312 which decided on a full-scale invasion. Having sacked Durham and received thousands of pounds for immunity from other towns, religious houses and local communities in the north, he withdrew in December. But there were still a number of English-held castles in Scotland, and he now set about reducing them. Perth was taken in January 1313, and those at Dumfries, Dalswinton, Buittle and Caerlaverock by the end of March. All were razed to the ground.

Outside Lothian, only two Scottish strongholds remained in English hands: Bothwell and Stirling. The latter, forming the gateway between the Lowlands and Highlands of Scotland, was by far the most important. While he himself was absent recapturing the Isle of Man, Bruce gave his brother Edward the task of forcing Stirling's surrender. Unfortunately, Edward Bruce was an impulsive man, a cavalry leader who abhorred the dull monotony of a long siege. So when Sir Philip Mowbray, the Governor of Stirling Castle, made the offer that, 'if by midsummer a year thence he was not rescued by battle, he would yield the castle freely,' Edward agreed. Given his brother Robert's avowed policy of never risking a pitched battle against the English, and the length of time he had allowed Mowbray for relief, it was an appalling error.

On his return from the Isle of Man, Robert the Bruce was apoplectic with fury, but he could not disavow the pledged troth of his brother. He made no secret of his displeasure, however, telling Edward that he had never 'heard so long a warning given to so mighty a King as the King of England'. 'We are so few against so many,' he continued. 'God may deal us our destiny right well but we are set in jeopardy to lose or win all at one throw.'

No sooner had Edward II heard of the terms of Mowbray's treaty with Edward Bruce than he began to marshal his forces. Faced with a common enemy, the Scots, his fractious barons agreed to meet him at Westminster Hall on 13 October 1313. Having apologized for the murder of Gaveston, Lancaster and his confederates received the King's pardon.

Everything was now in Edward II's favour. With England reunited, he could draw on five times as many troops as the Scots. Apart from its northern counties, his country had been untouched by recent wars and ranked as the most prosperous in Europe. Ireland and Wales were quiet, the papacy favourable and France, for the time being, friendly. Scotland, by comparison, had been ravaged by nine invasions and eighteen years of almost continual war. Its only hope was the military genius of its king and the enthusiasm of his subjects.

By June of 1314, at Berwick-upon-Tweed, Edward had assembled a 20,000-strong army – the greatest ever personally commanded in the field by a King of England. Foremost in rank and splendour were some 2,500 knights from all over Europe. Each was clad in chain mail overlaid by a surcoat with his armorial bearings, and accompanied by a squire and up to three men-at-arms. They came from as far away as Guienne and Germany, and included the Earls of Gloucester, Pembroke and Hereford, Sir Ralph de Monthermer, Sir Robert Clifford, and Sir Giles d'Argentan, described as the third best knight in Christendom.

Not among them were the sullen Earls of Lancaster, Warwick and Arundel, although in accordance with their feudal obligations they had sent their quota of cavalry and footmen. They excused their non-attendance by claiming that the consent of Parliament should have been obtained before Edward made war.

Supporting the horsemen were 3,000 Welsh longbowmen – so proficient that they could have five arrows airborne at the same time – and 15,000 foot soldiers, in quilted jackets and steel helmets, and armed with spears, swords and shields.

On 17 June, with just a week to go before the terms of the

Mowbray's challenge expired, Edward's huge army set out from Berwick. The King set a cracking pace, 'not as if he was leading an army to battle but as if he was on a pilgrimage to St James of Compostella', wrote Barbour, a contemporary chronicler. 'Brief were the halts for sleep, briefer still for food: hence horses, horsemen and infantry were worn out with toil and hunger.'

On 21 June, after a forced march of 22 miles through the dust and heat of a glorious midsummer's day, Edward's exhausted army reached Falkirk. There they bivouacked, with 10 miles and 36 hours to go before the expiry of the treaty.

The Scottish army, meanwhile, was waiting ahead in Torwood, a huge forest that lay either side of the Roman road from Edinburgh to Stirling, 5 miles north of Falkirk. Its vanguard, under the command of the Earl of Moray, numbered 500 men from the north of Scotland, including the towns of Inverness, Elgin and Nairn. The second division, led by Edward Bruce, was 1,000 strong and composed largely of men from Buchan, Angus, Lennox and Galloway. Of similar strength was the third division, drawn from the Lowland regions of Lanark, Renfrew and the Borders. Under the nominal command of a minor, James Stewart, the hereditary High Steward of Scotland, its actual commander was his cousin, James Douglas. Robert the Bruce led the fourth division of 2,000 men, mostly from the Western Highlands but including his own personal contingent from Carrick, Kyle and Cunningham.

To support these foot soldiers were 500 light horse under the Marischal, Sir Robert Keith, and a small company of archers from Ettrick Forest. In all, therefore, the Scottish army numbered a little over 5,000, about a quarter of the English host.

Possessing so few cavalry, Bruce based his battlefield tactics on the schiltron: the hedgehog of spears. Primarily defensive, it could nevertheless also act as a mobile battering ram. Months of careful training had ensured that the wild Highlanders who made up the majority of Bruce's army were able to manoeuvre with the level of discipline necessary for the schiltron to be effective.

As Bruce awaited his adversary, he skilfully selected the site of battle that best suited his tactics. Two miles to the north of Torwood, the road dipped down to the valley of the Bannock burn (or stream). Descending from hills to the west, the burn cut through wooded slopes and meadows, across a ford in the Roman road, down a steep gulley by the hamlet of Bannock, and away across the marshland to the north-east before emptying into the Firth of Forth. To the north of the burn were more natural obstacles: on the left of the road the royal forest of New Park; on the right, below a steep bank, the Carse of Balquiderock, a flat plateau of clayland enclosed by the Bannock burn and a subsidiary known as the Pelstream.

The English could not advance to the east across the marshland; nor could they make a detour to the west where Torwood and the New Park formed an unbroken forest. Their only route of advance, therefore, was either along the Roman road or to the east of Bannock where they could ford the burn and follow a small track along the escarpment at the Carse's edge.

Robert the Bruce therefore sent his vanguard to St Ninian's Kirk to watch the track along the Carse, with the divisions of Douglas and Edward Bruce ahead of it, to the left of the Roman road. He placed his own division to the right of the road, at the edge of the New Park. Keith's horsemen were in reserve a little further back at the foot of Gillie's Hill. On an L-shaped front, the army faced south-east down a gradual slope to the burn, with a clear view of the entries to both the New Park and the Carse.

At dawn the following morning, 23 June, the Scots attended

Mass; as it was the vigil of St John the Baptist, they consumed only bread and water. Once in position, the divisions were given a message from their King that if any were of faint heart they were free to depart. They responded with a great roar that they would conquer or die.

Edward, meanwhile, was approaching from Falkirk. At midday he reached Torwood and halted. He was met there by Mowbray who had made a wide detour from Stirling Castle to join him. A battle was unnecessary, Mowbray explained, because, under the laws of chivalry, Edward had fulfilled his obligation by bringing his army to within three leagues of its objective. Stirling could remain an English castle. But Edward would have none of it. He had not come all this way, he told Mowbray, to let his enemy slip away.

At the subsequent council of war, Mowbray explained that the English could not attack from the western flank as the paths through the forest had been barricaded. Edward therefore decided to make a frontal attack up the Roman road. His nephew, the Earl of Gloucester, would command the vanguard. If the Scots did not flee at the sight of such a force, Gloucester would scatter them with his heavy cavalry. At the same time, 600 knights under Sir Robert Clifford and Sir Henry Beaumont would advance along the track at the edge of the Carse to get behind the Scots and cut off their anticipated retreat.

At the head of Gloucester's column as it bunched to ford the burn was Sir Henry de Bohun. Emerging from trees on the north bank he could see a single rider across the open stretch of ground. It was Robert the Bruce. An axe in his hand and a golden circlet on his helmet, he was riding along the front rank of his division, half-hidden on the edge of the New Park.

Spotting de Bohun's arms on his surcoat, Bruce spurred forward; it was the de Bohuns who had been given his English lands by Edward I in 1306. The two charged, but at the last second Bruce swerved, rose up in his stirrups and brought his axe crashing down on de Bohun's head. So heavy was the blow that it sliced through de Bohun's helmet and penetrated his brain, splitting the axe handle in two.

Seeing this, the Highlanders climbed over their fieldworks and charged the English cavalry who were trying vainly to line up on the open ground below. Many had already fallen into pits, disguised with brushwood, that the Scotsmen had dug on either side of the road. Into this confusion raced the tartan-clad, garishly painted Highlanders. Within minutes, the panicked knights had fled. Careful not to allow a pursuit, Bruce ordered his men back to their positions.

Soon after, as his brother and senior officers were upbraiding him for putting his life in danger, he noticed the body of English horse led by Clifford and Beaumont, previously hidden by the bank along the Carse, appear from the direction of St Ninian's Kirk. 'A rose has fallen from your chaplet,' he shouted to Moray.

The chastened commander lost no time galloping back to his division and placing it on open ground ahead of the horsemen. Assailed on all sides, the hedge of spearmen held firm because the knights had been sent as a flying column and had no archers to assist them. In frustration, the cavalry hurled axes, swords and maces, but to little effect. Finally, seeing a gap between the horsemen, Moray drove his schiltron into it, splitting the English in two and sending some flying north to Stirling while the others returned to the main army.

Twice in one day, the cream of the English army had been repelled by foot soldiers, mainly because Edward, haughty and ignorant, had sent his knights in without the support of either infantry or bowmen. It seems a particularly strange omission given that many of his commanders had been present at Falkirk in 1298, where Wallace's defeat owed much to the devastation wrought by Edward I's archers.

It was late afternoon when news of the cavalry's second reverse reached Edward. He decided not to engage his disgruntled infantry, exhausted after the second forced march in two days, and put off continuing the attack until the following day. But this did not prevent defeatism from spreading through the ranks, and Edward was forced to send heralds around the army to explain that the engagements had been mere skirmishes and that the real battle was still to come.

To water his men and livestock, Edward led them down to where the Bannock burn flowed round the Carse. The foot soldiers and the supply column bivouacked on the southern bank. The cavalry were sent across to the hard clay of the Carse, where the open country that separated them from the Scottish positions was ideal for their deployment. Still Edward persisted in the absurd notion that his cavalry alone was capable of scattering the Scots.

Even this calculation, however, was based on the assumption that the Scots would remain on the defensive. In fact, on learning of Edward's new dispositions, Bruce resolved to move his men to the edge of the Carse. The English cavalry were enclosed on three sides by the Pelstream and the Bannock burn; by the early hours of 24 June the tide in the Firth of Forth would have risen, backing up these streams and making them impassable. With his men preventing their escape, the knights would be caught in a sack.

Before moving, Bruce addressed his commanders: 'Sirs, we have every reason to be confident of success for we have right on our side. Our enemies are moved only by desire for dominion but we are fighting for our lives, our children, our wives and the freedom of our country. And so I ask that with all your strength, without cowardice or alarm, you meet the foes whom you will first encounter so boldly that those behind them will tremble.'

Shortly after dawn on 24 June, the Scots celebrated Mass

and ate a light meal. Three divisions then moved off in echelon (i.e. one following the next, but offset so that each unit has a clear field of fire or manoeuvre before it): first Edward Bruce's, with its right flank protected by the Bannock burn; next, slightly to his left rear, Moray's; lastly, in the same manner, Douglas's. All were in schiltron formation. The King's division and the cavalry were kept in reserve on the lower slope of the New Park.

When Edward II saw the Scots advancing on foot over open ground he exclaimed: 'What, will yonder Scots fight?'

'Surely, sir,' replied Sir Ingram de Umfraville, 'but indeed this is the strangest sight I ever saw for Scotsmen to take on the whole might of England by giving battle on hard ground.'

As he spoke the Scots, now within a couple of hundred yards, knelt down to pray. 'They kneel for mercy,' cried the exultant English king.

'For mercy yes,' said Sir Ingram, 'but not from you. From God for their sins. These men will win all or die.'

'So be it,' said the King, ordering his trumpeters to sound the assembly.

The Earl of Gloucester was the first to mount. He was still smarting from the King's accusation of disloyalty the night before when he had suggested resting the men for twenty-four hours. Before his vanguard had a chance to range themselves behind him, he charged Edward Bruce's schiltron and was impaled on its spears. Many of his best knights, including Sir Robert Clifford, were similarly slain as, in ones and twos, they hurried to catch up with him.

When the bulk of the vanguard arrived, it was incapable of piercing the thicket of spears. Many riders were thrown as their horses were stabbed, leaving them defenceless on the ground. Then Moray's schiltron arrived on their flank, causing the survivors to wheel back and rejoin the main body of cavalry, which was trying to form up for action. This set off a stampede of wounded and riderless horses into the midst of the mustering squadrons.

Douglas now appeared on Moray's left and sealed off the only exit from the Carse; the tide had risen and the streams were difficult to cross. As the Scots pushed forward, the chaotic mass of cavalry were forced into an ever-diminishing space. Much of the English infantry had by now struggled across the water but they were effectively blocked from taking part in the battle by the knights ahead of them, while King Edward's archers could not fire for fear of hitting their own men.

Eventually, many of these archers were sent over to the north of the Pelstream from where they had a clear field of fire into the left of Douglas's division. But Bruce, from his vantage point on the slope of the New Park, saw the damage they were doing and sent Sir Robert Keith and his 500 light horsemen to disperse them. Those who were not cut down ran back among their own infantry, causing them, in turn, to flee.

In the middle of the Carse, the English knights were fighting with the desperation of trapped animals. 'The battle there was fiercest,' wrote Barbour, 'and so great was the spilling of blood that it stood in pools on the ground. There might be heard weapons striking on armour and knights and horses be seen tumbling on the ground and many a rich and splendid garment fouled roughly underfoot.'

Gradually, the English were giving ground. Seeing this, Bruce threw in his reserve division, particularly to bolster the left of Douglas's division which had been much reduced by the English bowmen. 'Press on, press on, they fail,' came the cry. With the added weight of the new arrivals, each man now pushed on the man in front so that the interlocking Scottish schiltrons, tipped with their 12-foot spears, ground forward like battering rams.

Assuming that all was lost, the Earl of Pembroke and Sir

Giles d'Argentan grabbed Edward II's bridle and, accompanied by some 500 knights of his bodyguard, struggled through the mêlée towards the now-ebbing Pelstream. Many Scotsmen tried to stop them and the King's shieldbearer was captured with the royal shield and seal. Edward's horse was stabbed, but it kept going until he was over the stream and a replacement was found. There Sir Giles took his leave, saying that he had never fled a battle; on returning to the fray, he was killed.

Edward continued on to Stirling Castle but its governor, Mowbray, would not admit him because, under the terms of the treaty, the castle would have to be surrendered and the King would become a prisoner. Instead, he lent Edward a local knight to guide him round the battle area and back to the safety of England.

With the departure of the Royal Standard, the English army began to disintegrate. Bruce speeded up the process by signalling his camp followers and late arrivals to advance from Gillies Hill. As this vast horde appeared over the crest, holding broadsheets for banners, the English mistook them for a second Scottish army and their slow retreat became a headlong flight.

'Never in the history of her wars had England suffered such a humiliation nor exhibited such helplessness in defeat,' wrote Ronald McNair Scott, author of *Robert the Bruce*. 'She had men and material enough to make an honourable stand. Her infantry had not even been engaged and many of her archers were among them: but not a leader emerged to rally them. Every armoured knight who had not been unhorsed or killed put spurs to his steed.'

Scores were drowned in the panic to cross the Bannock burn. The Earl of Pembroke, alone, kept his head. Returning to the field he gathered some thousands of his Welsh soldiers and led them south in orderly columns. Though harried along the route, most reached Carlisle safely. Leaderless, the majority of the infantry made for Stirling and took refuge beneath the castle. So numerous were they that Bruce kept his men in formation, convinced that the English would rally. But as he approached, the fugitives laid down their arms and Mowbray solemnly handed over the keys to the castle.

Thirty-four English barons and several hundred knights and squires were among the dead. Buried in consecrated ground, they were the lucky ones; the many foot soldiers also killed were piled in communal pits. The captives numbered almost a hundred barons and knights, including the Earl of Hereford and Sir Ingram de Umfraville. As well as the proceeds from their ransom, the Scots secured the whole of the English baggage train, said to stretch for 20 miles and to have been worth in excess of £200,000. Among the booty were gold and silver vessels, money chests, siege weapons, tents, wine, cattle, sheep, pigs and numerous spare warhorses.

Although the Scots lost just two knights, hundreds of pikemen were also killed. Nevertheless, by their sacrifice they had secured for their King the whole of Scotland with the exception of Berwick. Edward, on the other hand, had wiped out his father's gains in a single battle.

The ineffectual English king was finally deposed in January 1327 by a combination of his Queen, Isabella, and his leading barons. Replaced by his son, who became Edward III, he was imprisoned and then murdered in Berkeley Castle in the same year. The cause of death is reputed to have been a red-hot poker forced up his rectum. A suitable end, then, for a king who had little aptitude for politics and even less for war.

Sedan

The surrender of the main French field army at Sedan on 1 September 1870 – not to mention the capture of the Emperor

Napoleon III – effectively ended the Franco-Prussian War, although Paris held out until the following January. It was a catastrophe that owed as much to political interference as to military incompetence.

France had declared war on 19 July 1870, ostensibly because the King of Prussia refused to guarantee that there would be no repeat of the recent Hohenzollern candidature – when a cousin of his had been offered the Spanish throne, something to which France was bitterly opposed. In fact, Count Otto von Bismarck, Prussia's Chancellor, had cleverly goaded the indignant French to war by doctoring a telegram from his sovereign to appear more insulting than it really was. A successful conclusion to the conflict, he hoped, would enable him to complete the unification of Germany under Prussian dominance that had begun in 1866 with the defeat of Austria and the creation of the North German Confederation.

A crucial factor behind his decision to risk war was the poor state of the French Army. Smaller than Prussia's (thanks to ballot rather than universal conscription), its mobilization was slower, its artillery inferior and its general staff non-existent. As if all this were not enough, it had the added handicap of an amateur Commander-in-Chief – the ailing Emperor, Napoleon III. 'Is it true?' asked his cousin, Princess Mathilde, on being told that he intended to take personal command.

'Yes,' he replied.

'But you're not in a fit state to take it! You can't sit astride a horse! You can't even stand the shaking of a carriage! How will you get on when there is fighting?'

She received her answer within two weeks of mobilization, when a French army suffered crushing reverses at Froeschwiller and Spicheren near its eastern frontier. Split in two, the Army of the Rhine began to retreat in disarray along the entire Lorraine front, leaving the way open for the Prussians to advance into the heart of France.

Displaying uncharacteristic sense, Napoleon was all for handing over command to a professional and returning to Paris to take over the reins of government. But the capital was in an uproar and the Empress Eugénie, acting as temporary Regent while he was with his troops, advised him to stay away. 'Have you considered all the consequences which would follow from your return to Paris under the shadow of two reverses?' she inquired by telegram.

In fact, while not over eager for him to resume his political duties, public opinion was all for him relinquishing military command in favour of Marshal Achille Bazaine, the darling of the Left and a man who had risen from the ranks. But Bazaine's failing – other than his vulgar habits and inelegant figure – was his subordinate's mentality. 'Had Napoleon followed his inclination and returned to Paris,' wrote John Bierman, the Emperor's biographer, 'Bazaine might possibly have summoned up the personal resources to take charge; in the event, the Emperor's presence emasculated him.'

When Napoleon did finally leave Bazaine to his own devices in mid-August – in favour of joining the remnants of Marshal Edmé Patrice de MacMahon's army at Châlons-sur-Marne – it was too late. A Prussian army blocked Bazaine's line of retreat to Verdun and, after the bloody but inconclusive battle of Gravelotte (at which, surprisingly, the Prussians lost more men than the French), he returned with his troops to Metz and remained bottled up there for the rest of the war.

Napoleon, meanwhile, had arrived in Châlons to find a large, if ill-disciplined, force of 130,000 men, extensively rearmed and resupplied. On 17 August, he held a conference to discuss future plans although his personal participation was minimal. When one of those present suggested he should either be at the head of his troops or his government, he murmured, 'I seem to have abdicated.'

After much discussion, Napoleon meekly agreed to the

proposal put forward by his cousin, Plon-Plon: he would accompany MacMahon's army to a defensive position in front of Paris; but first General Louis Trochu, a liberal soldier popular with republicans, would prepare the political ground by returning to Paris as the newly appointed Military Governor.

When Trochu arrived to inform the Empress of Napoleon's intentions, she was not impressed. 'No, the Emperor will not return to Paris,' she said. 'Those who inspired the decisions you talk of are enemies. The Emperor would not enter Paris alive. The Army of Châlons will make its junction with the Army of Metz.'

Palikao, the recently appointed head of government and War Minister, was similarly dismissive, telling Trochu that he would never agree to MacMahon retiring on Paris.

Meanwhile, the Prussians were approaching Châlons and on 21 August MacMahon marched out with his troops – not west to Paris, nor east to engage the enemy, but 30 miles north-west to the fortified city of Rheims. Napoleon went with him.

Next day, Eugène Rouher, President of the Senate and Palikao's envoy, arrived at Rheims to persuade the Emperor not to return to Paris but to instruct MacMahon to march on Metz. When MacMahon was informed, he telegraphed Paris: 'How can I move towards Bazaine when I am in complete ignorance of his situation, and when I know nothing of his intentions?'

But no sooner had this message been dispatched than two others arrived. One, from Bazaine, dated 19 August, said that after allowing his men two or three days' rest he intended to break out of Metz to the north-west, before turning south-west towards Châlons; the other, from the Regency Council, addressed to the Emperor, read: 'Not to support Bazaine will have the most deplorable consequences in Paris. Faced by this disaster, it is doubtful whether the capital can be defended.'

Napoleon, therefore, was forced by his ministers to abandon the plan to fall back on Paris – the only sensible military course of action – in favour of linking up with Bazaine. To avoid the approaching Germans, MacMahon opted to march in a north-easterly direction in the hope of meeting Bazaine in the vicinity of Montmédy.

On 27 August, when he learnt that Bazaine had not moved from Metz and that two German armies were about to sever his lines of communication, MacMahon ordered his army to withdraw north. An urgent dispatch from Paris soon put a stop to this. 'If you abandon Bazaine,' Palikao warned, 'revolution will break out in Paris and you yourself will be attacked by the entire enemy forces.' He assured MacMahon that the latter was at least thirty-six hours ahead of his pursuers and that he had nothing in front of him 'but a feeble part of the forces which are blockading Metz'.

Accompanying the personal message was an official imperative: 'In the name of the Council of Ministers and the Privy Council, I require you to aid Bazaine.'

MacMahon could not defy a direct order from his political masters; cancelling his plans to escape to the north, he prepared to cross the River Meuse. Appealing to Napoleon was pointless – he was a broken man, without authority. 'If only I could die,' he had gasped to an aide the day before.

On 30 August, the Germans caught up with MacMahon's army at Beaumont near the Meuse. Supported by heavy artillery, they attacked at dawn and inflicted terrible casualties, although the French fought valiantly. By nightfall the majority of MacMahon's men had withdrawn in some disorder over the bridges at Mouzon and Villers. In the dark they continued on north up the Meuse valley towards the small fortress town of Sedan, just 7 miles from the Belgium frontier.

The following day, MacMahon deployed his demoralized troops around the town in a defensive triangle between the

Rivers Meuse, Floing and Givonne. He seemed blissfully unaware of the mortal danger his army was in and continued to believe that after a couple of days' rest he could continue his march – either east towards Metz, or west towards Mezières, where a new French corps was forming. He had no idea that 250,000 Germans with 500 cannon were about to make the escape of his 110,000 men impossible.

Field Marshal Helmuth von Moltke, the Chief of the Prussian General Staff, knew the truth. 'Now we have them in a mousetrap,' he told his officers that afternoon. General Auguste Ducrot, commanding the French I Corps, felt the desperate situation deserved a more earthy metaphor. 'Nous sommes dans un pot de chambre,' he exclaimed, gazing out over the encircling camp fires, 'et nous y serons bien emmerdés.' ('We're in a chamber pot and they're going to crap on us.')

Before dawn on 1 September, under the additional cover of fog, the Germans crossed the Meuse and attacked the bottom corner of the defensive triangle at Bazeilles. They were repulsed, but the fighting soon spread north along the line of the River Givonne. The French were soon suffering terribly from the accurate and sustained artillery fire. One of the first victims was MacMahon, wounded in the leg by a shell fragment while riding out to assess the situation. He named Ducrot as his successor.

Ducrot was quick to see that while the Givonne line was holding, the Germans were 'only amusing us there', and would soon attack along the left side of the triangle. He therefore ordered an immediate retreat through the north, and as yet unguarded, side of the triangle and thence west between a loop in the Meuse and the Belgian frontier.

Had Ducrot's instructions been carried out, a large part of the Army of Châlons might well have been saved before the net finally closed. That they were not was due to General Emmanuel Félix de Wimpffen, a former Military Governor of

Oran and only recently arrived from Paris. Before leaving the capital, Palikao had given him written authorization to take command if MacMahon was incapacitated. Now de Wimpffen produced the letter and, heartened by the successful action at Bazeilles, immediately countermanded the order to retreat.

'We need a victory,' he told the protesting Ducrot.

'You will be very lucky if by this evening you even have a retreat,' Ducrot replied.

'. . . we shall swallow Prussia at one gulp'

When war broke out with Prussia in 1870, the French were far from down-hearted. After all, they possessed the excellent breech-loading Chassepot rifle: accurate at 1,200 yards, it had twice the effective range of the Prussian Dreyse needle gun (also a primitive bolt-action breech-loader).

Napoleon III's troops also had the benefit of a machine-gun, the Mitrailleuse, which worked on the Gatling principle (a handle detonating its twenty-five barrels in turn). With a range of 2,000 yards and a rate of fire of 150 rounds a minute it was a deadly weapon. But so secret was its development that only a handful of troops had seen it before hostilities began. With no time to discover its best tactical use, the army deployed it as artillery rather than as an infantry weapon, and thereby reduced its effectiveness.

This was doubly unfortunate because the Prussians already had an artillery advantage with their excellent breech-loading Krupp guns. The French, by contrast, were still using the rifled, muzzle-loading pieces that had done good work in Italy in 1859. In 1867, having

cont. overleaf

watched Krupp cannon firing in Belgium, French officers reported on its superior range and accuracy. But nothing was done because so much money had already been spent on producing the Chassepot. A year later, Friedrich Krupp personally offered his guns for sale to the French government. Leboeuf, the French Minister of War, filed away the brochure and the reports with the comment: 'Rien à faire' ('Do nothing').

The other major Prussian asset was their General Staff. Set up during the Napoleonic Wars, this body of highly trained officers acted as the army's central nervous system and enabled it to mobilize and deploy with speed and efficiency. It had been working on the plan for an invasion of France since 1867.

By contrast, the French army had no General Staff and no detailed plan. Its ruling principle was 'On se débrouille' ('We'll muddle through') – jokingly referred to by the officer class as 'System D'. But this did nothing to dispel the preposterous optimism of the French generals before the fighting began.

'[They] all vouched for our victory . . .' recalled a bitter Empress Eugénie. 'I can still hear them telling me at [the imperial palace at] St Cloud, "Never has our army been in better condition, better equipped, in better fighting mettle! Our offensive across the Rhine will be so shattering that it will cut Germany in two and we shall swallow Prussia at one gulp." '

By now the sun had burned away the morning fog and the German artillery was pouring shells into the defensive triangle. Repeatedly, the French cavalry charged the German gun positions. But their gallant attempts were futile and they were thrown back by the supporting infantry with terrible losses. 'Ah, the brave fellows!' exclaimed the Prussian King Wilhelm I, watching with von Moltke from the safety of a hill south of the Meuse.

William Howard Russell, the veteran war correspondent of *The Times*, was observing the battle from the same hill. 'There must have been a hell of torture raging within that semicircle,' he wrote, 'in which the earth was torn asunder from all sides, with a real tempest of iron, hissing and screeching and bursting into the heavy masses at the hands of the unseen enemy.'

Into that storm rode Napoleon from the relative safety of the subprefecture in Sedan. Two of his aides were killed before he came across the new Commander-in-Chief. 'Your Majesty may be quite at ease,' said de Wimpffen, shells falling all around them. 'Within two hours I shall have driven your enemies into the Meuse.'

Napoleon rode on without comment. Shortly before 2pm, back in Sedan, he received a bizarre note from de Wimpffen: 'Sire, I have decided to force the line facing General Lebrun and General Ducrot rather than be taken prisoner in Sedan. I beg Your Majesty to place yourself in the midst of your soldiers, so that they may have the honour of opening a way for your retreat.'

Rightly, the Emperor ignored it, ordering instead that a white flag be raised on the ramparts. It was visible to Ducrot – his beaten corps huddling in and around the city – as he rode into the citadel. But Ducrot was more concerned with the 'indescribable' conditions that he found. 'The streets, the squares, the gates,' he recalled, 'were choked with carts, carriages, guns, the impedimenta and debris of a routed army. Bands of soldiers, without arms or knapsacks, streamed in every moment and hurried into the houses and churches. At the gates, many were trodden to death.'

Presenting himself to Napoleon, he was asked: 'Why is the firing going on? I have hoisted the white flag.' The Emperor then told him to write a note ordering all firing to cease. Ducrot did so, but refused to sign it on the grounds that only de Wimpffen had the authority.

A note was duly dispatched to de Wimpffen, but he refused to read it. The bearer then pointed out that the white flag had already been raised. 'No! No!' came the anguished reply. 'I will have no capitulation. Haul the flag down. I intend to continue the battle.'

He then hurried to Sedan to round up as many men as he could for a counter-attack. When he had gathered about 1,200 – using the duplicitous cry, 'Bazaine approaches! Bazaine approaches!' – he marched them south, but they were quickly routed.

Fortunately, Wilhelm I was just as anxious to end the slaughter, if only to save the lives of his own men, and he ordered every artillery piece to fire on Sedan as a means of convincing the enemy 'of the hopelessness of his situation'. It worked. With shells exploding in the garden, Napoleon ordered the white flag to be raised a second time. 'I obeyed a cruel but inexorable fate,' he wrote in exile two years later. 'My heart was broken, but my conscience was easy.'

At 6.30pm, General Reille handed Napoleon's letter of surrender to the Prussian King. 'My brother,' it read, 'having been unable to die among my troops, there is nothing left for me but to place my sword into the hands of Your Majesty.'

De Wimpffen then formally negotiated the capitulation of his army and wrote to Palikao: 'I came, I saw and I was defeated.'

More than 104,000 French troops were captured; a further 3,000 had been killed and and some 14,000 wounded. Prussian losses were 9,000 killed and wounded. The news reached Paris on the 3rd and a republic was proclaimed the following day. Fittingly enough, the two political figures who had done so much to bring about the defeat were among its first casualties:

Palikao was replaced by Trochu; the Empress Eugénie forced to flee to Britain, where she lived out her life as an exile.

St Valéry

On 12 June 1940, more than a week after the last British troops had been evacuated from Dunkirk, the 51st (Highland) Division was forced to surrender at the small Normandy seaside town of St Valéry-en-Caux. There had been many opportunities for the Highlanders to escape during the preceding days. But Churchill had insisted that they fought on in the hope of keeping France in the war. This was never likely to happen and they were sacrificed for nothing.

A Territorial formation, the Highland Division had made its name during the First World War when its kilted soldiers were known by the Germans as 'The Ladies from Hell'. Shortly after the outbreak of the 1939–45 war, the War Office decided that kilts were not suited to the demands of modern warfare and replaced them with battledress. The Highlanders retained their fearsome reputation.

They arrived in France during the 'Phoney War' in January 1940, only the second Territorial formation to join the British Expeditionary Force (BEF). Assigned to III Corps in positions opposite the Belgian border, much of their first two months was spent digging an anti-tank ditch.

In late April, the Highlanders were sent down to the Saar to do a tour of duty in front of the vaunted Maginot Line, the extensive fortifications along France's eastern frontiers. Single British brigades had been gaining valuable combat experience there since early December; but now General Lord Gort, VC, the British Commander-in-Chief, wanted to accelerate the 'battle-hardening' of the BEF by sending whole divisions. At first, the Regular 5th Division was chosen to go. But when Germany invaded Norway on 9 April, it was earmarked to take part in a relief expedition and the 'Fighting 51st' was sent instead.

The Highland Division was still occupying a forward sector of the Maginot Line when the Germans launched their long-awaited *Blitzkrieg* against the Low Countries – Operation Yellow – in the early hours of 10 May. Within five days, Holland had surrendered and no less than seven panzer divisions had pierced the French defensive line on the River Meuse beteen Dinant and Sedan. Foremost among them was Major-General Erwin Rommel's 7th Panzer Division, which by the evening of the 15th had advanced 20 miles beyond the Meuse.

'We are defeated,' said Paul Reynaud, the French Premier, in a telephone conversation with Winston Churchill during the morning of 15 May. 'We have lost the battle.'

Already, French morale had been dealt a mortal blow. As the military situation deteriorated, Churchill (who had been appointed Prime Minister of the Coalition Government on 10 May) was forced into an intricate game of political chess, with the 51st Division as a pawn, in a desperate attempt to keep France in the war.

On 20 May, the Highland Division was withdrawn from the Maginot Line as a prelude to returning it to the main body of the BEF. Such a contingency had been agreed with the French in the event of a German offensive, but it was now far too late to implement it. That same day, German panzers reached Abbeville and severed the lines of supply to the Allied left wing in Belgium.

The French General Maxime Weygand had by now replaced General Maurice Gamelin as the Allied Commander-in-Chief. But his plan to counter-attack north and regain contact with the armies trapped in Belgium never got off the ground, and on 25 May Gort made the bold but unauthorized decision to withdraw the BEF towards Dunkirk. The epic evacuation began two days later.

All the signs were that the French no longer had the stomach for a fight. Reynaud arrived in London on 25 May and told Churchill that he 'could hold out no hope that France had sufficient power of resistance'. If the Battle for France was lost, as it surely would be, no less a figure than Marshal Henri Philippe Pétain, the Deputy Premier, 'would speak in favour of an armistice'.

There was better news for the British Prime Minister on the 27th. First, his Chiefs of Staff reported that a German invasion could not succeed without naval or air superiority; then a telegram arrived from the British Ambassador to Washington, with an assurance from President Roosevelt that the United States would enter the war if the Allies really were *in extremis*.

Even so, Churchill was still anxious to keep France in the war for as long as possible. If her government sued for peace, Germany would gain control of her powerful navy, and this might make an invasion of Britain possible. Whereas if she fought on from her colonies or the New World, this would greatly assist the struggle against Hitler. Even if she only fought to the death on French soil, this would still provide Britain with a valuable breathing space in which to build up the country's defences. All Churchill's efforts, therefore, were geared towards convincing France that there was still hope.

On 31 May, he flew to Paris and told the French War Cabinet that he had every intention of sending more troops out to France as soon as they were properly equipped. Naturally, the two divisions still in France – the 51st and the 1st Armoured – would remain. He could not, however, spare any fighter planes because so many had been lost defending the Dunkirk beachhead.

According to Major-General Edward Spears, the British

liaison officer to the French government, Churchill already 'realized in his heart that the French were beaten, that they knew it, and were resigned to defeat'. His decision to leave British troops to fight on, therefore, can only be seen as a political gamble at very long odds. It was a risk he was prepared to take because he knew that aircraft and not troops were the short-term key to Britain's survival.

One senior officer who was convinced that these troops would be sacrificed for nothing was Lieutenant-General Alan Brooke, the commander of II Corps in the original BEF, and the man largely responsible for the success of the evacuation from France. On 2 June, just back from Dunkirk, he was called to a meeting with General Sir John Dill, the Chief of Imperial General Staff, and told that he would command the new BEF in France. 'I left his room,' wrote Brooke, 'with the clear conviction that what I was starting on was based purely on political requirements, and from what I had seen of the French up to date I had very great doubts as to any political advantages to be gained.'

By now, the Highland Division had been deployed on the lower Somme at the extreme left of the new French defensive line. It had been allotted an unprepared front of 18 miles, between the Channel coast and Pont-Rémy. The existing military practice was for one division to occupy no more than 4 miles of properly entrenched positions.

On 4 June, Major-General Fortune, the Highlanders' commander, directed an attack on the German bridgehead over the Somme at Abbeville. It was a fiasco. The French troops involved were late and performed poorly; the Highlanders fought gallantly but made little headway against numerous enemy machine-guns.

That same day, Lieutenant-General James Marshall-Cornwall (as he then was), the British liaison officer with the French Tenth Army (under which the Highland Division was fighting), sent a report to General Dill. It was quickly passed on to Churchill. In it, he suggested that a 'policy should be laid down covering the routes of withdrawal of the British Troops in the event of a German breakthrough further east'. He went on to warn that there was a danger of the Highland Division being 'driven into' the Le Havre peninsula and 'trapped there' as the bridges at Rouen were bound to be destroyed. His solution was for the French to agree to the preparation of alternative crossings over the Seine below Rouen.

Needless to say, nothing was done. Churchill was hardly going to make such a defeatist suggestion to the French when Weygand had just issued the stirring general order that there would be no retreat from the present line, that its defenders would throw back the enemy or die in the attempt.

But it was all rhetoric. The following day, 104 German divisions attacked the Somme-Aisne line between Luxembourg and the Channel. Opposing them were just fifty-one divisions: the Highland Division, the 1st Armoured Division (with only a third of its tanks), Beauman Division (a hotchpotch formation of untrained and poorly equipped line-of-communication troops), and forty-eight French divisions, many under strength and still being re-formed. The Germans, therefore, had a numerical superiority of two to one, an even greater imbalance of tanks and planes, and the psychological advantage of an unbroken run of successes.

The outcome was inevitable. Spread out over such a huge area, the Highlanders' front line was little more than a series of isolated posts. The Germans simply strolled through the gaps. Whole companies were cut off, later to surrender, and by the close of 6 June the survivors had been pushed back 8 miles to the River Bresle.

General Fortune was desperate. He had lost the equivalent of four infantry battalions in just three days' fighting and knew his exhausted men could not withstand such unequal odds for much longer. That afternoon, in a letter to General Altmayer, the commander of the French Tenth Army, he asked for his division to be relieved and placed in reserve. The inevitable reply was that 'there was no reserve formation available with which to relieve you'.

Fortunately, General Marshall-Cornwall was alive to the danger. In a telegram to General Dill, sent during the evening of the 6th, he pointed out that the division was 'hardly fit for more fighting' and that it might 'crack if seriously attacked'. 'If politically undesirable to withdraw all British troops from front line,' he concluded, 'I would urge that two more British Divisions with fighter support be sent urgently to France.'

One division, the 52nd (Lowland), was indeed on its way and Dill responded to Marshall-Cornwall's warning by instructing Brigadier Swayne, liaison officer with the French High Command, to secure a line of retreat for the 51st Division across the lower Seine, where the new BEF would concentrate. But that day Weygand had (falsely) accused the Highlanders of withdrawing 'without orders' and dubbed their commander 'Misfortune', and was in no mood to respond to this latest suggestion.

On 7 June, the Germans exerted little pressure on the Bresle line, but this was deliberate. If the left wing of the French Tenth Army remained static, this would give their panzers, making good ground further south, enough time to cut off its retreat. By the end of the day, Rommel's 7th Panzer Division had advanced more than 30 miles towards Rouen; the 5th Panzer was not far behind. In effect, the Tenth Army had been cut in half.

In a telegram to Dill, sent that evening, Marshall-Cornwall said that he had 'lost confidence' in the ability of the French 'to stop' the German drive to the Seine. He had asked General Altmayer to pull back the Highland Division and neighbouring French troops, to prevent them being trapped, but the army commander 'had refused to do so without orders from higher authority'. 'I suggest you come over and see Weygand immediately,' he concluded. 'Otherwise we shall have to evacuate 51st from Dieppe beach.'

Next morning, in desperation, Altmayer moved his headquarters over the Seine and virtually out of contact with his two corps. Marshall-Cornwall, who had not been informed in advance, angrily pressed him to order the withdrawal of French IX Corps (including the Highland Division) to the line of the River Béthune. Again Altmayer refused. Realizing further argument was useless, Marshall-Cornwall hastened to Weygand's headquarters near Paris.

From there he telephoned the War Office and warned that the Highland Division was 'in imminent danger of having its communications cut, and may have to be partially evacuated from the coast'. 'General Fortune,' he continued, 'should at once be released from French military control – certainly from that of Tenth Army – and given independent role of withdrawing his Division by successive lines to the Béthune and the Lower Seine.'

In fact, shortly before this conversation, the War Office had instructed General Howard-Vyse, its liaison officer, to inform Weygand that 'unless orders were given soon for the manoeuvre of the Allied wing' there was a 'grave risk of British troops being trapped'. This was a response to Marshall-Cornwall's message of the previous evening. But there was a codicil: Howard-Vyse was to represent in the 'strongest terms that evacuation between Dieppe and Le Havre cannot be contemplated'. For that might be seen by the French as another Dunkirk, another betrayal, and would give them just the excuse they needed to sue for peace.

A few minutes before midnight, possibly to ensure that an unauthorized evacuation did not take place, the War Office sent a signal to the Admiralty stating that 'the blocking of Dieppe could be carried out forthwith'. Despite Marshall-Cornwall's warning that the Highlanders might need 'to be partially evacuated from the coast', the War Office was determined to seal off its nearest point of escape.

Marshall-Cornwall, meanwhile, had managed to persuade Weygand to order IX Corps to retire over the Seine. But the problem now was time. The French, with their mainly horse-drawn transport, were never going to reach Rouen, or the ferry crossings below it, before the panzers. The largely motorized Highland Division might have done so, but it was under French orders and not in a position to desert its allies.

At a conference to discuss the move back to Rouen during the afternoon of the 8th, General Fortune was told by General Ihler, the commander of IX Corps, that it would take four days! If he was dismayed by this snail's pace he did not show it, although he must have known that the Germans would never afford them that sort of time. Nor did they, with Rommel's panzers reaching the outskirts of Rouen that evening.

The following morning, Fortune received word from the French admiral at Le Havre that the bridges at Rouen had been blown. However, the admiral had orders to arrange embarkation from his port 'if necessary'. Ihler broke down on hearing the news, leaving Fortune to assume overall command. He decided to make for Le Havre, 60 miles down the coast.

Why he chose Le Havre rather than Dieppe, less than a day's journey away, is something of a mystery. Swinburn, his Chief of Staff, wrote that Dieppe was 'dismissed as impractical owing to the . . . destruction of the harbour at an earlier date'. In fact, according to a naval report, the harbour was still open, although there was the possibility of mines outside. It was not until the 10th that the Royal Navy carried out the War Office's instructions and sank blockships outside the harbour entrance. Furthermore, there was the possibility of embarkation from the beaches either side of the town (where the Canadians would land two years later).

Yet Fortune had not received permission to embark from Dieppe – and he was only too aware of the political consequences of taking such a unilateral decision. The British government would not intervene because it was terrified of upsetting the French; indeed, its actions tend to indicate a more sinister agenda. During the afternoon of 9 June, when it already knew that German troops had reached Rouen, the War Office reminded Fortune that his 'aim should be to break south of the Seine on the axis Dieppe–Rouen'.

Knowing that such a course of action was impossible, Fortune replied: 'Withdrawing with French Forces to Havre as crossings at Rouen blown.' In fact, he had arranged for part of his division – 'Ark Force' – to race on ahead and take up a line 20 miles east of Le Havre. The rest would follow as quickly as the marching French would allow.

It was not fast enough. During the afternoon of 10 June, just hours after Ark Force had passed that way, Rommel's panzers reached the coast at Les Petites Dalles. By evening they had moved east and taken the crossings over the River Durdent. For the main body of the Highland Division there would be no evacuation from Le Havre. Shortly before 11pm, General Fortune cabled the War Office: 'Think possible that in this rapidly changing situation I might ask you to embark as much personnel as possible of my Division between St Valéry and mouth of River Durdent.'

The plan was altered soon after when word arrived that the Germans held the Durdent. The Highlanders and the French would now try to hold a perimeter round the small seaport of St Valéry-en-Caux, a natural break in the high chalk cliffs of the Côte d'Albâtre that stretch from the Somme to Le Havre.

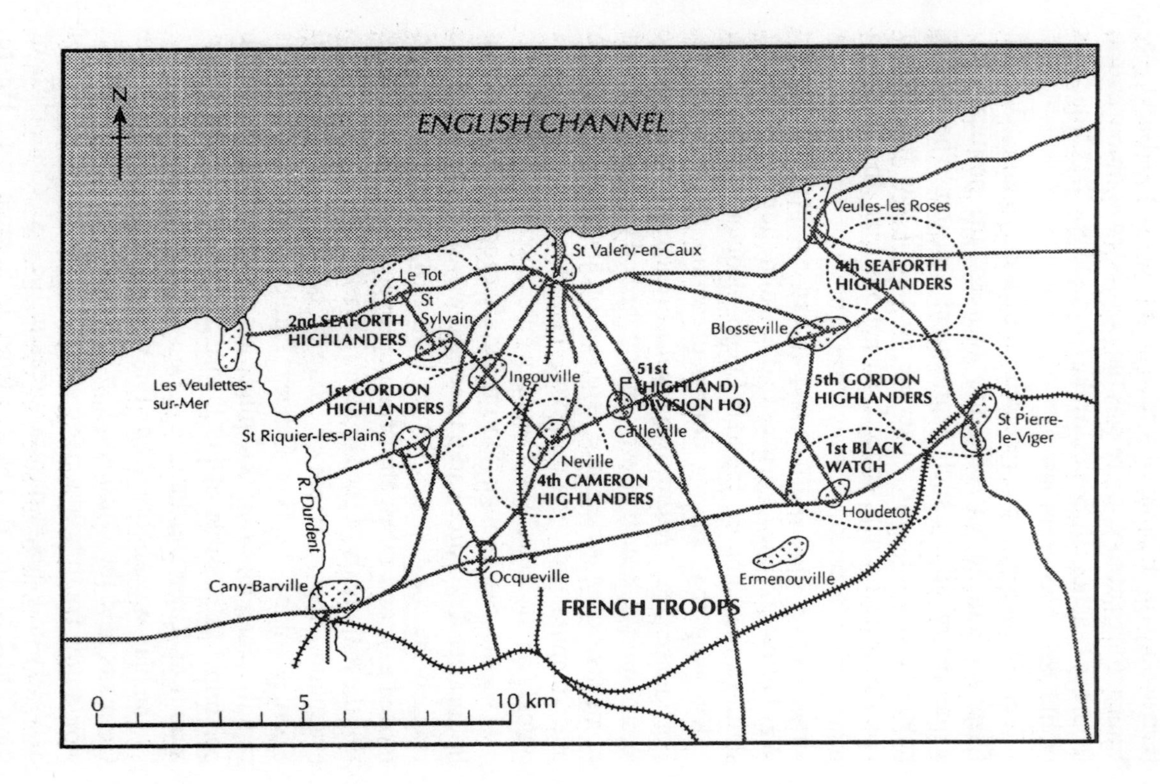
N
ENGLISH CHANNEL
Veules-les Roses
St Valery-en-Caux
Le Tot
St
Sylvain
4th SEAFORTH
HIGHLANDERS
2nd SEAFORTH
HIGHLANDERS
Blosseville
Les Veulettes-
sur-Mer
Ingouville
51st
(HIGHLAND)
DIVISION HQ)
5th GORDON
HIGHLANDERS
1st GORDON
HIGHLANDERS
Calleville
St Pierre-
le-Viger
St Riquier-les-Plains
Neville
1st BLACK
WATCH
4th CAMERON
HIGHLANDERS
R. Durdent
Houdetot
Ocqueville
Ermenouville
Cany-Barville
FRENCH TROOPS
0
5
10 km

At 8.45am the following morning, 11 June, Fortune again cabled London: 'Intend to embark whole force to-night Tuesday provided sufficient ships and boat transport are available.'

The War Office's absurd response was to send a copy of an order from Weygand to Ihler, issued the day before, for IX Corps to withdraw over the Seine below Rouen. The message ended by reminding Fortune of 'the importance of acting in strict conformity with any orders IX Corps commander may issue'.

Despite the fact that it knew the Highland Division was trapped on the coast, with its only hope of escape the sea, the War Office was still urging its commander to obey an order that would lead to its certain destruction. Can we infer from this, then, that Churchill and the British government actually needed to sacrifice the Highlanders to confirm their determination to stand by the French to the end? It seems possible. Certainly, the loss of the Highland Division, so well known to the French from the First World War, would go some way to negating the bad feeling caused by Dunkirk and the subsequent holding back of British fighters.

But if this was Churchill's intention, Fortune was not going to comply meekly. 'Physical impossibility Corps Commander [General Ihler] approach Seine. In same boat as me,' he replied at 10am. 'Air Control of the enemy area round my bridgehead essential.'

Fortunately, one man was still working to save the Highlanders: Admiral Sir William James, Commander-in-Chief, Portsmouth. On 8 June, he had been ordered by the Admiralty to prepare Operation Cycle, the embarkation of British line–of-communication troops from the Le Havre peninsula. The following morning, having gathered together more than 200 merchant vessels of all sizes, he dispatched them to Le Havre with a destroyer escort. But after discovering the plight of the

Highland Division during a visit to Le Havre on the 10th, he ordered the whole armada to set sail for St Valéry. The first to arrive were two destroyers and a transport ship. They made contact with Fortune's Naval Liaison Officer the following morning, 11 June, and agreed to begin the evacuation that night.

Sadly, it was one day too late. Pinning the Highland Division (12,000 men) and the four fragmented French divisions of IX Corps (another 12,000 men) into a bridgehead just 7 miles wide and 5 deep were 5 German divisions – 2 panzer, 2 infantry and 1 motorized – numbering more than 50,000 men. Around 2pm on 11 June, they attacked. Despite stout resistance from the 2nd Seaforth Highlanders and the 1st Gordon Highlanders, Rommel's 7th Panzer Division quickly broke through the outer defences and occupied the cliffs to the west of the town. With German guns commanding both the harbour and the beaches, any attempt to embark large numbers of troops from St Valéry was bound to end in a bloodbath.

Shortly after 5pm, Rommel sent a captured French soldier into St Valéry with a message for its defenders: surrender by 9pm or face an all-out bombardment of the town. Back came the response: the British had no intention of capitulating. Incredibly, General Fortune still had every intention of embarking as many men as he could that night. His optimism seemed to be borne out by the arrival of a signal from Le Havre at 6pm: the French admiral commanding there had at last given permission for the embarkation. Only now was Admiral James authorized by the War Office to attempt a rescue. At 6.15pm, he signalled the officer commanding the rescue armada: the evacuation would begin that evening.

Brief verbal orders were sent out to all Highland Division units, giving times of embarkation from 10.30pm onwards. At least two battalions holding the perimeter – the 2nd Seaforths and a portion of the 1st Black Watch – never received them;

others, like the 5th Gordons, did, but far too late to give them any chance of making their embarkation slot.

It would not have helped even if they had been on time, however. Shortly after midnight, the first British ships arrived off St Valéry. Captain Warren, commanding the destroyer HMS *Codrington*, could see that 'the place was burning fiercely and a lot of machine-gun and artillery fire was being directed on the beach'. As it was clearly 'quite impracticable as a place of evacuation', he directed all incoming boats to the small coastal village of Veules-les-Roses, 4 miles to the east.

Between 2 and 9 o'clock in the morning, Admiral James's armada picked up more than 1,300 British and 900 French troops from the beach at Veules. Many had slipped away from their posts without orders. Significantly, few were from the Highland infantry regiments manning the outer perimeter.

General Fortune was never informed that the evacuation of St Valéry had been called off, and his men waited on the seafront until 3am. Afraid that they would be dangerously exposed when daylight came, he ordered them to withdraw to woods south of the town. There they were joined by a number of late-arriving perimeter troops.

As dawn approached, Fortune had still not conceded defeat. If his men could retake the heights on the west of the town, and hold out for one more day, they might still be rescued the following night. But General Ihler was not convinced and responded by asking Fortune to forward a telegram to Weygand, via London, informing him of the surrender of IX Corps. Fortune refused to send it until he had consulted with his senior commanders.

When they, understandably, showed little enthusiasm for his doomed plan, he decided to go ahead regardless. The 4th Seaforths were given the most onerous task: to drive Rommel's panzers off the western cliffs. A little after 8am, as the attack was being prepared, a white flag was seen fluttering from the church steeple. Minutes after he had ordered it to be taken down, General Fortune received an order from Ihler to cease fire. He refused, signalling the War Office that he would not comply while there was 'a possibility of evacuating by boat any of my division later'.

At 10am, as German shells began exploding next to the building housing his headquarters, he accepted that further resistance was futile and gave the order to surrender. It was not a moment too soon for the men of the 4th Seaforths who were forming up on the start line ready to attack.

In all, 10,000 members of the Highland Division were taken prisoner at St Valéry. Another 1,000 or so had been captured on the Somme. Fatal casualties were more than 1,000, with four times that number wounded. And all because Churchill's government was determined not to give the French another excuse – as if they needed one – to seek an armistice. It made no difference. Five days after the capture of the Highlanders, the new French government under Marshal Pétain sued for peace.

'It has always been abundantly clear to me,' wrote Fortune's Intelligence Officer, Captain Ian Campbell (the future Duke of Argyll), 'that no division has ever been more uselessly sacrificed. It could have got away a good week before but the powers that be – and owing I think to very faulty information – had come to the conclusion that there was a capacity for resistance in France which was not actually there.'

'Scotland's Pride'

The slate was finally wiped clean on 2 September, 1944, when the re-formed 51st (Highland) Division – which had risen like a Phoenix from the ashes of its predecessor – liberated St Valéry from the Germans. Field Marshal Montgomery, the British commander in

Normandy, had generously given it the opportunity to repay the four-year debt owed to the original Highland Division by changing his order of battle.

As the infantry marched into the main square behind five pipers, they were given a tremendous welcome by the French townsfolk, some of whom were wearing Highland kilts that they had kept hidden for four years. At the head of the 5th Black Watch was Lieutenant-Colonel Bill Bradford, a captain in the original division whose memorable escape had taken him to Gibraltar, via Marseilles and Algiers. It was fitting that he should be asked to lay the wreath at the memorial service for the men who had died in 1940.

Another successful escaper was the divisional commander, Major-General Thomas Rennie, also a major in 1940. Nearing St Valéry, he had reminded his troops of the fate of their comrades: 'That magnificent Division was sacrificed to keep the French in the war. True to Highland tradition, the Division remained to the last with the remnants of our French Allies, although it was within its capacity to withdraw and embark at Le Havre.'

On 3 September, after the massed bands of the Division had Beaten the Retreat, Rennie once again addressed his men:

> Here, at St Valéry on the 12th June, 1940, a portion of the Highland Division, including its Headquarters, 152 and 153 Brigades, was captured by a large German force.
>
> That Highland Division was Scotland's pride; and its loss, and with it the magnificent men drawn from practically every town, village and croft in

cont. overleaf

Scotland, was a great blow . . . It has been our task to avenge the fate of our less fortunate comrades and that we have nearly accomplished . . . We have lived up to the great traditions of the 51st and of Scotland.

North Africa: 1940–1

The fighting that raged in North Africa from September 1940 to April 1941, between the British on the one hand and the Italians (and later the Germans) on the other, saw a quite extraordinary sequence of unlikely victories and lost opportunities. Much of the responsibility must be borne by two politicians – Mussolini and Churchill – who could not resist interfering.

On 10 June 1940, with the defeat of France certain, the fascist dictator Benito Mussolini had brought Italy into the war on Germany's side in the hope of gaining some of the spoils of victory. In particular, he wanted to supplant Britain as the dominant power in North Africa and the Mediterranean (Italy already had colonies in Libya and, since its invasion in 1936, Abyssinia). But he should have taken more notice of a remark made by Sir Percy Lorraine, the British Ambassador in Rome, on being informed by Count Galeazzo Ciano, the Italian Foreign Minister (and Mussolini's son-in-law), that their two countries were at war. 'I have the honour to remind Your Excellency,' replied Sir Percy, 'that England is not in the habit of losing her wars.'

At least Ciano seemed to have some inkling of what Italy was getting herself into. As Mussolini proclaimed war to an unenthusiastic Roman crowd from the balcony of the Palazzo Venezia, Ciano moaned: 'I am sad, very sad. The adventure begins. May God help Italy.'

To be fair to Mussolini, Britain was in a desperately weak position. While much of her army had been rescued from Dunkirk, the troops had left behind most of their weapons, transport and equipment, and in that weakened state faced imminent invasion by the Germans. She was hardly in a position to do much if the Italian armies in Libya and East Africa chose to attack the tiny British garrisons that guarded Egypt and the Sudan.

Furthermore, Italy's entry into the war made the Mediterranean too dangerous for troop convoys, which had to be sent by the long route around the Cape instead. One small batch of 7,000 troops, ready to leave Britain in May, did not reach Egypt until the end of August. Even their arrival did little to diminish the massive Italian troop superiority in the theatre. While General Sir Archibald Wavell, the British Commander-in-Chief, Middle East, had less than 50,000 troops at his disposal, the Italians had more than half a million – although about 100,000 were native levies.

On the North African front alone, the Italians had more than 300,000 men under Marshal Rodolfo Graziani in Libya; Egypt, on the other hand, housed just 36,000 British, New Zealand and Indian troops. But this did not mean that the British were prepared to stand on the defensive in the Western Desert. Just days after the declaration of war, a mobile column of the 7th Armoured Division – the original 'Desert Rats' – captured the forts of Capuzzo and Maddalena, on the border between Libya and Egypt. 'At small cost we have inflicted casualties wherever enemy forces encountered, capturing 25 Italian officers and 500 ORs [other ranks],' Wavell signalled London on 22 June. 'Early attacks completely surprised enemy.'

Of course, there was no question of holding these forts. With such a numerical inferiority, Wavell's only option was to harass the Italians by a series of hit-and-run operations. It was with such guerrilla tactics in mind that he formed the Long Range Desert Group at the end of June.

Irked by these minor British successes, Mussolini became desperate for a propaganda victory. On 15 July, at his insistence, the Italian High Command ordered Graziani to prepare an offensive against Egypt (which, although a base for British forces, was technically neutral). Graziani, however, was far from enthusiastic. On 8 August, after weeks of procrastination, he was called to an interview with Ciano in Rome. Intended to stiffen the Marshal's resolve, it seemed to have little effect. 'Our present preparations are far from perfect,' complained Graziani, adding: 'We move towards a defeat which, in the desert, must inevitably develop into a rapid and total disaster.'

He might not have been so despondent if he had known about Wavell's supply difficulties. By the end of July, more than 200 of the 306 tanks possessed by Lieutenant-General Richard O'Connor's Western Desert Force – the precursor of the Eighth Army – were under repair. On 7 August, after a conversation with Wavell in London, Anthony Eden, the War Secretary, described the deficiencies as 'shocking'.

But thanks to Graziani's hesitation, there was still time to send a large supply convoy of tanks, guns and spare parts round the Cape. It finally arrived at the Suez Canal on 5 September, by which time an Italian army from Abyssinia (now Ethiopia) had overrun British Somaliland. The Italians had not, however, had it all their own way. Despite outnumbering the British troops by more than five to one, they were held up for four days at the Tug Argan Pass. When the British were finally evacuated by sea on 18 August, they had inflicted more than 2,000 casualties and suffered barely 250.

Far from making Mussolini more cautious, the Somaliland campaign merely whetted his appetite. On 29 August, having run out of patience, he personally ordered Graziani to launch an attack from Libya by 10 September. Yet still Graziani delayed. On 7 September, Mussolini told him to attack in two days' time or resign. At last, reluctantly, the Marshal gave the necessary orders. 'Never,' wrote Ciano, 'has a military operation been undertaken so much against the will of the commander.'

Even then, the attack took place four days late because some units became lost on their way to the assembly point, while Graziani suffered renewed doubts on hearing reports of 'massive British armoured forces' gathering to oppose him. In fact, Wavell's strategy was to withdraw all forces eastwards across the Western Desert to Mersa Matruh, 120 miles inside the frontier and 200 miles west of the Nile Delta.

On 13 September, therefore, the six divisions of the invading army crossed the frontier into Egypt unopposed. However, after covering just 60 miles, Graziani decided to dig in around the coastal town of Sidi Barrani. Even worse, he dispersed his troops in a chain of fortified camps that were too widely separated to support one another. Ciano was horrified, noting in his diary entry for 2 October: 'Graziani insists that we shall have to wait quite a long time, at least until the end of November, in order to complete preparations for renewed advance.'

But the delay was good news for the British. Two days before the Italian advance, General Wavell had asked his Chief of Staff to consider plans for an invasion of Libya, noting: 'We may . . . hope to be dealing with a somewhat dispirited and not very formidable Italian, and to be able to take a certain degree of risk.' Graziani's half-hearted invasion seemed to confirm this. It also gave the British enough time to rush out reinforcements, including three armoured regiments, aboard fast merchant ships.

But Wavell would not be hurried, even after Italy declared war on Greece on 28 October and there was a danger that some of his meagre force would be sent to that theatre. 'It seems now is the time to take risks and strike,' Churchill cabled on 14 November. 'Operation is in preparation,' Wavell replied two days later, 'but not possible to execute this month as originally hoped. Now working to date about end first week December unless enemy moves meanwhile.'

There was little danger of that and Wavell was able to complete his plans without hindrance from the Italians – although Churchill continued to press for an early start. However, Wavell never intended the attack to become a full-blown offensive. It was, he signalled to the Prime Minister on 6 December, 'designed as a raid only'. He explained: 'We are greatly outnumbered on ground and in air, have to move 75 miles over desert and attack enemy who has fortified himself for three months.'

Operation Compass was designed to take advantage of Graziani's dispersed defences around Sidi Barrani. The string of fortified camps stretched from Maktila on the coast to Sofafi, 50 miles inland, and behind this the Italians were deployed in depth. But the forward camps were not mutually supporting and there was a big gap between Sofafi and the next nearest group of camps at Nibeiwa. The plan was for General O'Connor's troops to surge through this gap at night before turning north to roll up the remaining forward defences. With a troop inferiority of 30,000 to the Italians' 80,000, the British were relying on their armoured advantage: 275 tanks to 120. The fifty heavily armoured – if undergunned – Matilda tanks of the 7th Royal Tank Regiment (RTR), impervious to most Italian anti-tank weapons, would prove particularly decisive.

On the night of 7 December, O'Connor's troops began the long approach over the desert from Mersa Matruh. The following night, as a mixed force with dummy tanks took up decoy positions in front of Maktila and Nibeiwa, the 4th

Indian Infantry and 7th Armoured Divisions passed through the gap in the Italian defences. Next morning, after a preliminary bombardment from the east, the Indians, spearheaded by Matildas of the 7th RTR, assaulted Nibeiwa from the rear. Gaining complete surprise, they destroyed twenty-three Italian tanks parked outside the camp. Within two hours, the camp itself had been taken, and with it 4,000 prisoners.

The 4th Indian Division then headed north and, again led by the Matildas, successfully stormed the camps of Tummar West and East before day was out. Meanwhile, the 7th Armoured had passed to the north of the camps at Sofafi and continued its drive in a north-westerly direction to the coastal town of Buq-Buq, thereby cutting off the enemy's line of retreat.

Sidi Barrani fell during the evening of the 10th after an attack from both flanks, supported by two of the 7th Armoured's tank regiments. Next day, the 7th's reserve brigade reached the coast beyond Buq-Buq and trapped 14,000 retreating Italians. In all, O'Connor's men had captured a staggering 40,000 troops, 237 guns and 73 tanks in just 3 days. Their only casualties were a meagre 624 killed, wounded and missing.

The remnants of Graziani's demoralized army streamed across the border into Libya and took refuge in the coastal fortress of Bardia. They were quickly followed up by the 7th Armoured Division, which cut the road to the west. Unfortunately, there was no infantry division available to take advantage of the situation because Wavell, ever cautious, had recalled the 4th Indian Division on the 12th and sent it to bolster the defence of Sudan.

The British C-in-C's lack of ambition is clear in the signal sent to London that day: 'Mobile column of Armoured Division will advance to Sollum-Capuzzo and endeavour to get across Tobruk road and cut off Bardia. If Bardia falls, though I think this unlikely, I have instructed O'Connor that he can push on towards Tobruk up to the limit of endurance of vehicles and men.'

It took three weeks for the 6th Australian Division to arrive from Palestine. This gave the Italian commander at Bardia, General 'Electric Whiskers' Berganzoli, plenty of time to put his 18-mile defensive perimeter in order. He assured Mussolini that his troops would fight to the end, and he had reason to be confident given that he had the equivalent of four infantry divisions (about 45,000 men). However, it did not seem to occur to him, any more than it had occurred to Graziani, that the best form of defence was attack, particularly when your opponent is numerically weaker.

On 3 January 1941, the assault on Bardia began with a violent artillery bombardment. Once two lanes had been cleared through the anti-tank ditch, wire and minefield in the eastern sector, the Australians poured through, led by the last twenty-five serviceable Matildas. By noon, Italians were surrendering in their thousands, although pockets held out for another two days. In honour of this victory, Eden coined a new version of Churchill's famous eulogy to the Battle of Britain pilots: 'Never has so much been surrendered by so many to so few.'

With each successive virtually bloodless victory – the Australians lost just 456 men killed and wounded – Wavell seemed to grow in confidence. On the day Bardia fell he ordered plans to be prepared for the capture of Tobruk and then Benghazi, the capital of Cyrenaica (Libya comprised two provinces, Cyrenaica and Tripolitania). As ever, he was being urged on by Churchill who, on 6 January, warned: 'Time is short. I cannot believe Hitler will not intervene soon . . .'

Yet all the while, Churchill was considering the removal of precious forces from the Middle East to aid Greece; for though the Italian campaign there had proved to be another fiasco, it looked as if the Germans were about to take over. 'Destruction of Greece would eclipse victories you have gained in Libya,' cabled Churchill to Wavell on 10 January. 'Nothing must hamper capture of Tobruk but thereafter all operations are subordinated to aiding Greece.'

Tobruk fell to the Australians on 22 January, with the remaining sixteen Matildas again playing a prominent part. A further 30,000 prisoners, 236 guns and 87 tanks were taken. By now, the Greek leadership had refused British military assistance, fearing it would provoke the Germans to attack, and Wavell had permission to continue on to Benghazi. But as Churchill had feared, Mussolini had by now accepted Hitler's offer of reinforcements for North Africa. The first German troops were due to arrive by the middle of February.

Time was therefore of the essence. Despite the fact that the 7th Armoured Division was down to just 50 cruiser and 95 light tanks, O'Connor pressed on. Finding the Italians in strong positions on the coastal road at Derna, 160 miles east of Benghazi, O'Connor planned an ambitious left hook. Not only would he bypass Derna, but also Benghazi, thereby trapping the last of Graziani's original army of invasion. But first he had to wait for more tanks and supplies, due to arrive on 7 February.

Four days before that date, however, he received air reconnaissance reports that the Italians were preparing to abandon Benghazi and retreat to the El Agheila bottleneck, from where they could block the route from Cyrenaica into Tripolitania. If he was going to stop them he had to move fast.

On the morning of 4 February, the depleted 7th Armoured Division, led by the armoured cars of the 11th Hussars, set out on one of the most daring escapades of the desert war. With two days' rations and barely enough petrol, they took just 33 hours to cover 170 miles of some of the most rugged terrain in

North Africa. They were just in time. Two hours after reaching the coastal road near Beda Fomm, the 11th Hussars, a battalion of the Rifle Brigade and some supporting artillery intercepted the advance elements of the Benghazi garrison. Although numbering less than 2,000 men, the British had the element of surprise and for three hours they kept the Italians at bay until the first armour arrived.

Next morning, the enemy's main column arrived. It was escorted by more than 100 new cruiser tanks, whereas the British had only 29. But the Italian armour arrived in dribs and drabs, and kept near the road, while the British manoeuvred into favourable 'hull-down' positions – that is, with their hulls protected by the lie of the land and only the turret exposed to fire. By nightfall, sixty Italian tanks had been knocked out and thousands of soldiers had surrendered. A final attempt was made to break out the following morning, led by sixteen tanks, but it was checked by the Rifle Brigade. Numbering less than 3,000 men, the British had taken 20,000 prisoners, 216 guns and 120 tanks.

In under two months, and at a cost of just 2,000 casualties, two Commonwealth divisions had advanced 500 miles and routed an enemy force five times as large. They had captured 130,000 Italians, and taken or destroyed nearly 400 tanks and over 800 guns. With the total annihilation of Graziani's army, the British now had a free run through the El Agheila bottleneck to Tripoli, Mussolini's last foothold in North Africa. But just as O'Connor was preparing to deliver the *coup de grâce*, Churchill intervened.

The Greek leader, General Ioannis Metaxas, had died on 29 January and his successor, Koryzis, had told the British Ambassador that he welcomed military assistance. Churchill was delighted and on 12 February ordered Wavell to halt the advance, transfer to Greece 'the fighting portion of the Army which has hitherto defended Egypt and make every plan for sending and reinforcing it to the limit'. If Greece, with British aid, could hold off the Germans for a few months, Churchill explained, the 'chances of Turkish intervention [on the Allied side] will be favoured'.

As it happened, Churchill's division of Wavell's meagre resources brought about disaster in North Africa as well as Greece. The first contingent of a force that would eventually number 50,000 arrived at Salonika on 7 March. The Germans invaded a month later and within weeks had overrun the country, forcing the British into yet another ignominious evacuation. They left behind 12,000 men, all their tanks and most of their equipment.

Back in North Africa, meanwhile, the tide was turning. On 6 February, the now Lieutenant-General Rommel was summoned to Hitler's headquarters and given command of the newly formed Afrika Korps, comprising two mechanized divisions, the 5th Light and 15th Panzer. Eight days later, the advance elements of the 5th Light Division – a reconnaissance battalion and an anti-tank battalion – began to arrive at Tripoli. They were immediately rushed up to the forward Italian positions, along with some dummy tanks that Rommel had had built on Volkswagen chassis. The divisional tank regiment did not arrive until 11 March.

When the British still failed to appear, Rommel decided to take the initiative. On 24 March, his reconnaissance battalion easily took the strategically vital but thinly held El Agheila. Led by fifty tanks, and with two new Italian divisions bringing up the rear, he decided to push on over two fronts.

It helped that the opposition was so inexperienced. Sent back to Egypt at the end of February to rest and refit, the 7th Armoured Division had been replaced by part of the newly arrived 2nd Armoured Division (the remainder had been sent to Greece). The excellent 6th Australian Division had also gone to Greece, its place taken by the 9th Australian which was short of equipment and training. O'Connor, too, had been temporarily relieved by General Neame, an untried commander.

Knocked off balance by Rommel's sudden advance, the British fell back in confusion and Benghazi was evacuated on 3 April. O'Connor returned to advise Neame, but both were captured when their unescorted car ran into German motor-cycle troops on the night of the 6th. Next day, the commander of the 6th Armoured Division was captured at Mechili, along with a newly arrived motorized brigade and other units. The encircling force had made itself appear more powerful than it really was by using trucks to raise dust clouds and imitate tanks. By 11 April, with Rommel's offensive just over two weeks old, most British troops had withdrawn across the Egyptian frontier. Only a small force holding Tobruk re-mained in Cyrenaica.

Like Mussolini before him, Churchill's intervention in the North African campaign had proved fatal to his country's military interests. Without it, O'Connor would probably have ejected the Italians from Libya before Rommel's forces could have made a difference. Even the Germans were convinced of this. 'We could not understand at the time,' wrote General Walter Warlimont, a member of Hitler's staff, 'why the British did not exploit the difficulties in Cyrenaica by pushing on to Tripoli. There was nothing to check them. The few Italian troops who remained there were panic-stricken, and expected the British tanks to appear at any moment.'

It would take another two years of heavy fighting, and many more lives, before the Axis powers were finally cleared from North Africa. 'The price to be paid,' wrote Liddell Hart, 'for forfeiting the golden opportunity of February 1941 was heavy.'

Stalingrad

On 31 January 1943, after a bitter siege of more than two months, Field Marshal Friedrich von Paulus surrendered his encircled German Sixth Army to the Russians at Stalingrad. It was fitting that this catastrophic defeat marked the beginning of the end for Hitler because he had interfered with his generals' running of the campaign from the outset.

Germany's invasion of Russia – Operation Barbarossa – had begun on 21 June 1941. At first, the armoured spearheads carried all before them and hundreds of thousands of Russians were taken prisoner. Gradually, however, the Germans became the victims of their own success: the further they drove into Russia, the longer their lines of communication became, the worse the weather, the stiffer the resistance of the enemy. As winter set in, making further advances impossible, they had reached the Crimea in the south, Leningrad in the north and come within an ace of taking Moscow.

Yet, with unlimited space on which to fall back, the total defeat of the Russians was unlikely. The situation became even more ominous for the Germans in December 1941, when the Japanese attacked Pearl Harbor and America entered the fray. Hitler was now faced with the spectre of a war on two fronts. To avoid this it was essential to undermine Russia's fighting capability as quickly as possible. But how? The German High Command was all for occupying Moscow and establishing a defensive line along the upper and middle Volga. This would have had the effect of crippling Russian communications, blocking the ingress of Lend-Lease supplies from Archangel, and cutting off the Russian armies west of the Urals from the resources of Asiatic Russia.

But Hitler had other ideas. Ignoring his generals, he took the advice of his leading industrialists and economic advisers, who warned him that the Reich would collapse unless the oilfields of the Caucasus were seized (with no oil resources of its own. Shortage of fuel for military operations and for industry was a major problem for Germany throughout the war). He decided, therefore, to advance from Batum on the Black Sea to Baku on the Caspian. If successful, he would strengthen his own military potential whilst weakening that of the Russians. To cover such an operation, a defensive front would need to be established along the River Don from Voronezh to Stalingrad.

The plan, however, had serious drawbacks. With insufficient forces to carry out both operations, the 360-mile defensive front would be lightly held and vulnerable to counter-attack. If the Russians broke through and took Rostov, the German forces in the Caucasus would be cut off from their base. Furthermore, since the Volga was ice-bound and therefore easily crossed for up to half the year, Stalingrad was of only minor strategic importance.

Hitler was unswayed by such arguments, preferring to exaggerate the capability of his own troops while underestimating that of the enemy. According to Colonel-General Franz Halder, his Chief of Staff, when it was pointed out to the Führer that Stalin still had 1,500,000 men in the region north of Stalingrad, not to mention 500,000 in the Caucasus, 'he flew at the man who was reading with clenched fists and foam in the corners of his mouth, and forbade him to read such idiotic twaddle.' As far as Halder was concerned, Hitler's 'decisions had ceased to have anything in common with the principles of strategy . . . They were the product of a violent nature which acknowledged no bounds to possibility, and which made the wish the father of the deed.'

In truth, Hitler's forces on the Eastern Front were less formidable than they might have appeared. Although more numerous than in 1941, they were undoubtedly weaker. Of a total of 232 divisions, 61 were from satellite nations like Romania, Italy, Finland and Hungary – generally of low morale and poorly equipped. Even the German infantry divisions were undermanned, with six battalions instead of nine; while only ten of the original twenty armoured divisions were back at full tank strength, because priority at the time had been given to submarine production.

The Russian army, thanks to its enormous losses in 1941, was composed mainly of Asiatics of low intelligence but huge natural tenacity and endurance. All able-bodied males were liable to be conscripted straight into front-line units – learning on the job, as it were – causing Field Marshal Fritz von Manstein to compare the dynamic of this army to that of the revolutionary armies of France, 'a combination of fanaticism and terror'. Crucially, the Russians also had a technical advantage with the simple but effective T34 tank, superior to any German equivalent until the introduction of the Panther and Tiger tanks in 1943.

In April, Hitler reorganized his forces by dividing Army Group South into the two army groups that would take part in the operation: A and B. His final plan was for the Fourth Panzer Army of Army Group B to advance on Voronezh, but not to occupy it. Followed by the Sixth Army, it would then wheel south-east down the right bank of the Don towards Stalingrad. Simultaneously, the Second German, Second Hungarian, Eighth Italian, and Third Romanian Armies would take over the defence of the river west of Stalingrad. Covered by this manoeuvre, Army Group A would advance toward the lower Don near Rostov, with the First Panzer Army leading. Once over the Don it would be joined by the Seventeenth Army and, ultimately, by the Eleventh Army once the latter had completed the conquest of the Crimea.

On 8 May 1942, General von Manstein's Eleventh Army reopened the Crimean campaign. It was a spectacular success:

Kerch falling on the 15th and the fortress of Sevastopol, after a siege of just one month, in late June. More than 250,000 prisoners were taken. Meanwhile, to relieve the pressure, the Russians had launched a massive offensive of their own either side of Kharkov on 12 May. But after initial gains, they were counter-attacked by Army Group B and eventually surrounded near Izyum on 26 May. The Germans captured a further 240,000 men, 2,026 guns and 1,249 tanks.

The Stalingrad campaign was therefore delayed and did not begin until 28 June. Within a week, the forward elements of the Fourth Panzer and Second Armies had reached Voronezh, while the Sixth Army was making good progress to the south. The Fourth Panzer Army then began to move south down the Don while the sixth Army shadowed it. A couple of days later, Army Group A began its drive towards the Caucasus.

On 17 July, the Russians evacuated Voroshilovgrad, north of Rostov, and fell back to the south-east, closely followed by the Seventeenth Army, while the First Panzer Army crossed the Donetz at Kamiensk. It was now that Hitler made a fateful alteration in his battle plans. Worried that the First Panzer Army would not be strong enough to force the lower Don, he ordered General Hoth to move the bulk of his Fourth Panzer Army to support it, leaving the von Paulus's Sixth Army to take Stalingrad alone. Halder objected: without the Fourth Panzer spearhead, the pace of the Sixth Army would slacken, giving the Russians time to put Stalingrad into a state of defence. But Hitler would not change his mind.

Events, for a time, seemed to justify this change of plan. On 19 July, Hoth's troops gained a bridgehead over the lower Don at Tsymlanskaya. Three days later, Rostov fell to the First Panzer Army and the Seventeenth Army crossed the Don in four places. The Sixth Army, too, was making good ground, and on 24 July reached the bank of the Don to the west of Stalingrad.

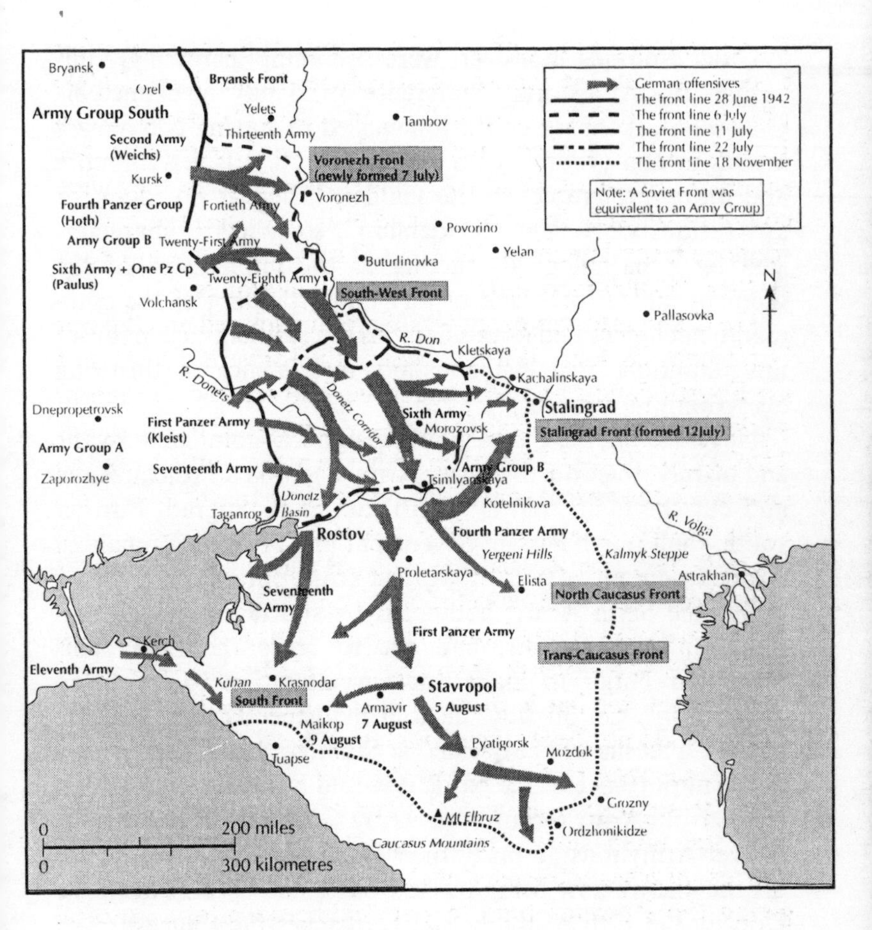

German offensives
The front line 28 June 1942
The front line 6 July
The front line 11 July
The front line 22 July
The front line 18 November
Note: A Soviet Front was equivalent to an Army Group
Bryansk
Orel
Army Group South
Bryansk Front
Yelets
Tambov
Second Army (Weichs)
Thirteenth Army
Voronezh Front (newly formed 7 July)
Kursk
Voronezh
Fourth Panzer Group (Hoth)
Fortieth Army
Povorino
Army Group B
Twenty-First Army
Yelan
Buturlinovka
Sixth Army + One Pz Cp (Paulus)
Twenty-Eighth Army
South-West Front
Volchansk
Pallasovka
R. Don
Kletskaya
Kachalinskaya
R. Donets
Donetz Corridor
Stalingrad
Dnepropetrovsk
First Panzer Army (Kleist)
Sixth Army
Morozovsk
Stalingrad Front (formed 12July)
Army Group A
Zaporozhye
Seventeenth Army
Army Group B
Tsimlyanskaya
Kotelnikova
Donetz Basin
Taganrog
Rostov
Fourth Panzer Army
R. Volga
Yergeni Hills
Kalmyk Steppe
Proletarskaya
Astrakhan
Elista
North Caucasus Front
Seventeenth Army
First Panzer Army
Trans-Caucasus Front
Kerch
Eleventh Army
Kuban
Krasnodar
Stavropol
South Front
Armavir
5 August
Maikop
7 August
9 August
Tuapse
Pyatigorsk
Mozdok
Grozny
Mt Elbruz
Ordzhonikidze
Caucasus Mountains
0
200 miles
0
300 kilometres
N

Hitler's moods, however, were becoming increasingly unpredictable. At a stormy conference on 23 July, he blamed his General Staff for the confusion caused by having two panzer armies in and around Rostov. 'The situation is getting more and more intolerable,' wrote Halder. 'There is no room for any serious work. This "leadership", so-called, is characterized by a pathological reacting to the impressions of the moment and a total lack of any understanding of the command machinery and its possibilities (i.e., Hitler is incapable of understanding that his constant interference is throwing everything in disorder).'

Nevertheless, the Caucasus campaign was going well. By the end of July, the Fourth Panzer Army had taken Proletarskaya and cut the main railway north-east to Stalingrad. Further south, the Fourth Panzer, Seventeenth and Fourth Romanian Armies were all advancing against scant opposition.

But the Sixth Army, hampered by shortages of fuel, ammunition and armour, had yet to cross the Don. This prompted Hitler to make another alteration to his troop dispositions. On 30 July, he told his generals 'that the fate of the Caucasus will be decided at Stalingrad, and that in view of the importance of the battle it would be necessary to divert forces from Army Group A to Army Group B . . . that the 1st Panzer Army must at once wheel south and south-west to cut off the enemy now being pushed back step by step from the Don by the 17th Army, before he reaches the Caucasus.'

Halder noted: 'This is rankest nonsense. This enemy is running for dear life and will be in the northern foothills of the Caucasus a good piece ahead of our armour and then we are going to have another unhealthy congestion of forces before the enemy front.'

On 1 August, in line with the first item of Hitler's instructions, the Fourth Panzer Army was returned to Army Group B and ordered to advance along the Novorossisk–Stalingrad railway. At first it encountered little resistance and reached Kotelnikovo – 100 miles from Stalingrad – within two days. But from there on the opposition intensified and by 9 August it had been forced on to the defensive.

The Sixth Army, meanwhile, had taken Kalach and crossed the Don. On 23 August, it reached the Volga and occupied Stalingrad's northern outskirts. Soon after it closed the gap between the Don and Volga Rivers, north of the city, and on 2 September established contact with Hoth and his Fourth Panzer Army at Kotelnikovo. Ten days later, von Paulus was ordered to take Stalingrad by storm.

By now, Army Group A had seized the Maikop oilfields and reached the edge of the Caucasus Mountains; but it could go no further. Part of the reason, according to its new commander, General Ewald von Kleist, was a lack of fuel. 'But,' he added, 'that was not the ultimate cause of the failure. We could still have reached our goal if my forces had not been drawn away bit by bit to help the attack on Stalingrad. Besides part of my motorized troops, I had to give up the whole of my flak [anti-aircraft] corps and all my air force except the reconnaissance squadrons.'

Part of the reason for Hitler's preoccupation with taking Stalingrad was symbolic in that the city bore the name of his deadly adversary. This of course ignored the many difficulties that such a task presented. An extended industrial city of half a million inhabitants, Stalingrad stretched along 18 miles of the west bank of the Volga. East of the northern and southern quarters of the city, the river flowed through several channels created by islands. If the Germans could bridge them, and invest the city on its eastern side, then it was only a matter of time before its garrison would be starved into submission. Instead, Hitler insisted on the city being taken by *coup de main*.

The battle began on 15 September, and after a week's hard

fighting the Germans had reached the city centre. By the 27th they had penetrated the factory district in the northern suburbs and taken the 'Iron Heights'. Two days later, however, they were ejected by a strong Russian counter-attack. Reinforcements were brought up – thereby weakening the armies guarding the flanks – and on 4 October, supported by tanks and bombers, the attack resumed. After a further ten days of ferocious street-fighting, having made little progress, the Sixth Army was exhausted.

Hitler now ordered a change of tactics: the assaults would cease while the city was systematically razed by shells and bombs. But this simply replaced buildings with rubble; the latter being more easily defended. Now the ground gained could be measured in yards. As the fighting became increasingly desperate, both over and under ground, the Germans took to describing it as the *Rattenkrieg* ('rat war').

On 9 November, Hitler insisted that 'not one square yard of ground will be given up'. The last general attack took place three days later, when the riverbank in the south of the city was reached. By mid-November, as well as fighting for Stalingrad, the Sixth Army was holding the Don–Volga gap and a stretch of the Don up to Kletskaya. From there to the German Second Army at Voronezh, the front was held by the Third Romanian, Eighth Italian and Second Hungarian Armies. Below Stalingrad the Fourth Romanian Army was guarding the Ergeni Hills, while to its rear the Fourth Panzer Army was refitting at Kotelnikovo.

It was now that Marshal Georgi Zhukov, the Russian commander, chose to unleash two massive counter-offensives either side of Stalingrad. On 19 November, General Konstantin Rokossovsky's army, spearheaded by three armoured and four cavalry corps, with twenty-one infantry divisions following up, smashed through the Romanians holding the Don between Kletskaya and Serafimovitch. His left-wing, however, was repulsed by the Sixth Army as it tried to advance through the Don–Volga gap. The following day, General Yeremenko's smaller army of two armoured corps and nine infantry divisions scattered the Romanians in the Ergeni Hills.

On 22 November, Yeremenko's forces linked up with Rokossovsky's near Kalach. The Sixth Army – numbering 270,000 men, 70,000 of them non-combatants – was surrounded. But all was not lost. With the Russians so disorganized, von Paulus could have broken out 'at any time during the following week', wrote the military historian Major-General J.F.C. Fuller.

The sceptical Halder had been relieved of his post in late September. General Zeitzler, his replacement, now urged Hitler to order von Paulus to fight his way out. But Hitler would have none of it. Reich Marshal Hermann Goering had assured him that the Luftwaffe could airlift at least 500 tons of supplies a day to ensure the Sixth Army's survival. On 24 November, therefore, the Führer told von Paulus to 'hedgehog' himself in and await relief.

The task of cutting a corridor through to Stalingrad was given to von Manstein. But not with the intention of enabling von Paulus to withdraw; rather, to re-establish the Stalingrad front. To achieve this unlikely objective, von Manstein was given the newly formed Army Group Don, comprising the Sixth, Fourth Panzer, and Third and Fourth Romanian Armies.

Von Manstein's plan, heavily influenced by Hitler, was to push the Fourth Panzer Army up the Kotelnikovo–Stalingrad railway, defeat Yeremenko, and then turn on Rokossovsky's left flank as von Paulus made a simultaneous attack from Stalingrad. He would then use General Hollidt's group – German reinforcements and the remnants of the Third Romanian Army – to strike against Rokossovsky's right.

The attack began on 12 September. Nine days later, the forward elements of the Fourth Panzer Army were within 30 miles of von Paulus's 'hedgehog'. Further north, however, a Russian army under General Nikolai Vatutin had broken through Hollidt's group and the neighbouring Eighth Italian Army. With the Fourth Panzer Army in danger of being cut off, von Manstein ordered von Paulus to break out within twenty-four hours. Impossible, he replied, his tanks had only enough fuel to drive 20 miles. His generals then urged him to abandon his transport; but without a direct order from Hitler he would not budge. By Christmas Day, Army Group Don was in full retreat and the fate of the Sixth Army was sealed.

On 29 December, Hitler was informed that the Sixth Army was rapidly running out of supplies. Contrary to Goering's assurances, the Luftwaffe had only managed to deliver an average of 200 tons a day (at a cost of 246 transport planes in less than a month). Yet the actual daily requirement was 700 tons, and these figures would double when all the reserve supplies were exhausted. Despite this hardship, Hitler was insistent: the Sixth Army had to hold out until the spring.

That same day, Zeitzler managed to persuade Hitler to withdraw Army Group A from the Caucasus. With von Manstein struggling to keep a corridor open, it made it just in time, crossing the Don on 22 January 1943. By the end of January the Don front had collapsed and a 200-mile gap separated von Manstein's left flank at Voroshilovgrad from the German Second Army at Voronezh in the north.

Conditions at Stalingrad, meanwhile, were desperate. With air supplies down to under 100 tons a day (they would cease entirely after 21 January), there was a shortage of food, medical supplies, fuel and ammunition. Typhus and dysentery killed thousands; the cold – with temperatures as low as –28 degrees – many more.

On 8 January, Rokossovsky called on the Germans to surrender. Those who did so would be properly treated, he promised. Von Paulus was anxious to comply. There was little hope of relief and he knew that the Russians were massing for a major assault. He therefore sent a signal to Hitler requesting 'freedom of action'. The Führer's response was that von Paulus had to hold on until February, when three panzer divisions, on their way from France, would attempt to break through. It was a pipe dream.

Two days later, Rokossovsky ordered a general assault. The Pitomnik airfield, 14 miles west of the city centre, fell on the 14th, prompting von Paulus to report that his men could hold out no longer. 'Capitulation is impossible,' replied Hitler. 'The Sixth Army will do its historic duty at Stalingrad until the last man, in order to make possible the reconstruction of the Eastern Front.'

The last German airfield was overrun on the 25th, reducing the Sixth Army to two isolated pockets, one in the north of the city, one in the south. Von Paulus was in the latter, with his headquarters in the basement of the Univermag department store. Such was the shortage of food that he now forbade the issuing of rations to the 30,000 sick and wounded.

On 30 January, the anniversary of Hitler's accession to power, von Paulus was simply going through the motions. 'The swastika flag is still flying high above Stalingrad,' he reported. 'May our battle be an example to the present and coming generations, that they must never capitulate even in a hopeless situation, for then Germany will emerge victorious.'

The following day, to stiffen von Paulus's new resolve, Hitler promoted him to field marshal – but it made no difference. That afternoon, the Sixth Army headquarters sent its final radio message: 'The Russians are before our bunker. We are destroying the station.' Von Paulus then surrendered, although the troops in the northern pocket –

General Strecker's VI Corps – continued to resist for a further forty-eight hours.

Hitler's initial reaction was to compare the Sixth Army with the Three Hundred at Thermopylae: it had shown the world 'the true spirit of National Socialist Germany and its loyalty to the Führer'. Later he raged against von Paulus's treachery. 'This hurts me so much,' he was heard to say, 'because the heroism of so many soldiers is nullified by one single characterless weakling.'

The fate of the Sixth Army was unique in that its casualties were absolute: 110,000 were already dead; 34,000 wounded and sick had been evacuated by air; the remaining 91,000 soldiers and 40,000 non-combatants were taken prisoner. However, only about 5,000 would ever see Germany again. Within six weeks, more than 50,000 were said to have died of starvation and disease in the brutal Russian prisoner-of-war camps.

Other losses included 500 transport planes and the equivalent of six months' production of tanks and vehicles, three months' production of artillery, and two months' production of small arms and mortars. And all because Hitler, a gifted amateur, but one prone to monstrous flights of fancy, had taken over the day-to-day running of the war from his generals.

Goose Green

The two-day battle for the settlements of Darwin and Goose Green – 28–29 May, 1982 – during the Falklands War ended in a crushing British victory. But it was an unnecessary sideshow, fought for political rather than strategic factors, and left the 2nd Battalion, the Parachute Regiment (2 Para) with fifteen men dead (plus two from other units), including its adjutant and its commanding officer.

Hostilities had erupted two months earlier when, in an attempt to divert domestic political unrest by settling a long-running territorial dispute, Argentina's ruling junta had ordered the invasion of the Falkland Islands ('Islas Malvinas') – a British possession since 1833. In the small hours of 2 April, Argentinian commandos landed near Port Stanley, the islands' capital, and after a brief but fierce firefight the tiny Royal Marine garrison was ordered to surrender by the civil governor, Rex Hunt.

Within days, a large naval Task Force had sailed from Portsmouth to retake the islands. It numbered two aircraft carriers, nine destroyers and frigates, numerous support ships, and the troopship SS *Canberra* with 3 (Commando) Brigade on board. By the time it reached the Falklands, in late April, it had been reinforced by the 2nd and 3rd Battalions, the Parachute Regiment and seven more destroyers and frigates, besides more support vessels.

At first, all went well. On 25 April, a mixed force of SAS, SBS (Special Boat Service) and Royal Marines, backed by Royal Navy vessels, regained possession of South Georgia, some way to the south-west of the main archipelago. A week later, the British nuclear submarine HMS *Conqueror* sank the Argentinian cruiser *General Belgrano*, despite the fact that it was some 40 miles south-west of the British-imposed Total Exclusion Zone extending for 200 miles around the islands. On 4 May, however, the Argentinians struck back when an Exocet missile, fired from a French-built Super Etendard aircraft, struck and crippled the destroyer HMS *Sheffield*, which ultimately sank.

The plan to retake the islands went ahead none the less, and on 21 May an unopposed beachhead was established at San Carlos Water on the west of the main island, East Falkland. But the day was not without mishap. HMS *Ardent*, a frigate guarding the landing, was sunk by bombs and two further

Royal Navy ships were badly damaged. On 24 May, a second frigate – HMS *Antelope* – was lost, and the following day the destroyer HMS *Coventry* and the container ship *Atlantic Conveyor* were sunk, the latter by an Exocet. The loss of the container ship was a particularly heavy blow, as with it were lost six out of ten Wessex and three out of four giant Chinook helicopters, upon which the plan to break out of the beachhead depended.

This was the final straw for the British government, already fearful that public support for the war was slipping and that the United Nations might insist on a ceasefire before its troops reached Port Stanley. A morale-boosting victory was essential, and the sooner the better. On 26 May, at the behest of the War Cabinet, Brigadier Julian Thompson, commanding 3 Commando Brigade (and temporarily commanding the ground forces in the Falklands until Major-General Jeremy Moore arrived), was ordered to engage the Argentinians at the first opportunity.

The obvious target was the enemy garrison straddling the twin settlements of Darwin and Goose Green, 13 miles south of San Carlos, in the narrow isthmus that connects Lafonia to East Falkland. A successful attack would have the added bonus of securing the nearby airfield, from which Argentinian Pucara ground-attack planes operated, and the release of 120 civilians being held prisoner in the Goose Green community hall.

But Thompson objected on the grounds that it was strategically irrelevant: once Port Stanley fell, Goose Green would also fall. His preference was to leave a small force to cover the isthmus, while pushing on with the rest of his men towards Mount Kent, the jumping-off point for an assault on the capital. No problem, replied Major-General Richard Trant, his military superior in London, but Goose Green had to be taken first. 'If ever there was a politicians' battle,' wrote Max

Hastings, co-author of *The Battle for the Falklands*, 'then Goose Green was to be it.'

Thompson summoned his commanders and gave them their orders. 45 Commando, Royal Marines and 3 Para would spearhead the break-out from the beachhead on foot, advancing towards Port Stanley across a wintry landscape of frozen peat and icy marshes. At the same time, the 450 men of 2 Para would attack the Darwin–Goose Green settlement, defended, according to intelligence reports, by a weak battalion of around 400 demoralized conscripts. The Paras would be supported by just three 105mm guns, all that could be lifted into position with the helicopters available.

Commanding 2 Para was the fiercely ambitious Lieutenant-Colonel Herbert 'H' Jones. Born into a landed West Country family and educated at Eton, he had done most of his soldiering with the Devon and Dorset Regiment before joining the Paras in 1980. Aged forty-two, this was his first time in action and he was determined to make the most of it. Informed by a senior gunner that it might help Thompson to postpone the attack by insisting on more fire support, 'H' replied: 'I'm not delaying anything.'

To support the attack, Major Chris Keeble, Jones's second-in-command, asked for some Scorpion and Scimitar light tanks. The request was refused on the grounds that petrol was in short supply and the staff did not think the tanks could cover the difficult terrain. The same reasons ruled out 'BVs' – Volvo-tracked transport vehicles – forcing the Paras to leave behind six of their eight heavy mortars. Covering fire would be provided by the guns of the frigate *Arrow*, while Sea Harrier strikes could be called in after dawn on the 28th.

At 3am on 27 May, cold and exhausted after having covered the 9 miles from Sussex Mountain, overlooking San Carlos Water, in full battle order, the Paras took refuge in Camilla

Creek House, 4 miles to the north of the nearest Argentinian positions. The plan was to rest up during the day and attack that night. Unfortunately, things started to go wrong soon after daylight when the men heard a BBC World Service broadcast which stated 'a parachute battalion is poised and ready to assault Darwin and Goose Green'.

Jones was enraged that such vital information had been leaked and told the BBC correspondent accompanying the battalion that he would sue the Defence Secretary if any of his men died in the forthcoming battle. He then deployed the battalion across a wide defensive area to meet the expected enemy air or artillery attacks. None came, but his men still had to spend another uncomfortable day out in the open. A far more ominous consequence of the ill-timed BBC broadcast was the decision by Major-General Mario Menendez, the Argentinian commander in the Falklands, to send reinforcements to the Darwin–Goose Green garrison from his strategic reserve. They arrived the following morning.

The attack by 2 Para began in the early hours of 28 May. A Company was the first into action, engaging the defenders of Burntside House, at the eastern entrance to the isthmus. After a brief but fierce firefight, the Argentinians fled, leaving behind two of their dead and four terrified but unharmed civilians. By 5.30am, unhindered by further resistance, A Company had reached its second objective, Coronation Point overlooking Darwin. The company commander, Major Dair Farrar-Hockley, son of General Sir Anthony Farrar-Hockley, the Colonel Commandant of the Paras, now radioed for permission to press on. But Jones, several hundred yards further back, asked him to wait so that he could observe the situation in person. For more than half an hour, as the last of the darkness disappeared, A Company held its ground. Finally, Jones arrived and the company could move again.

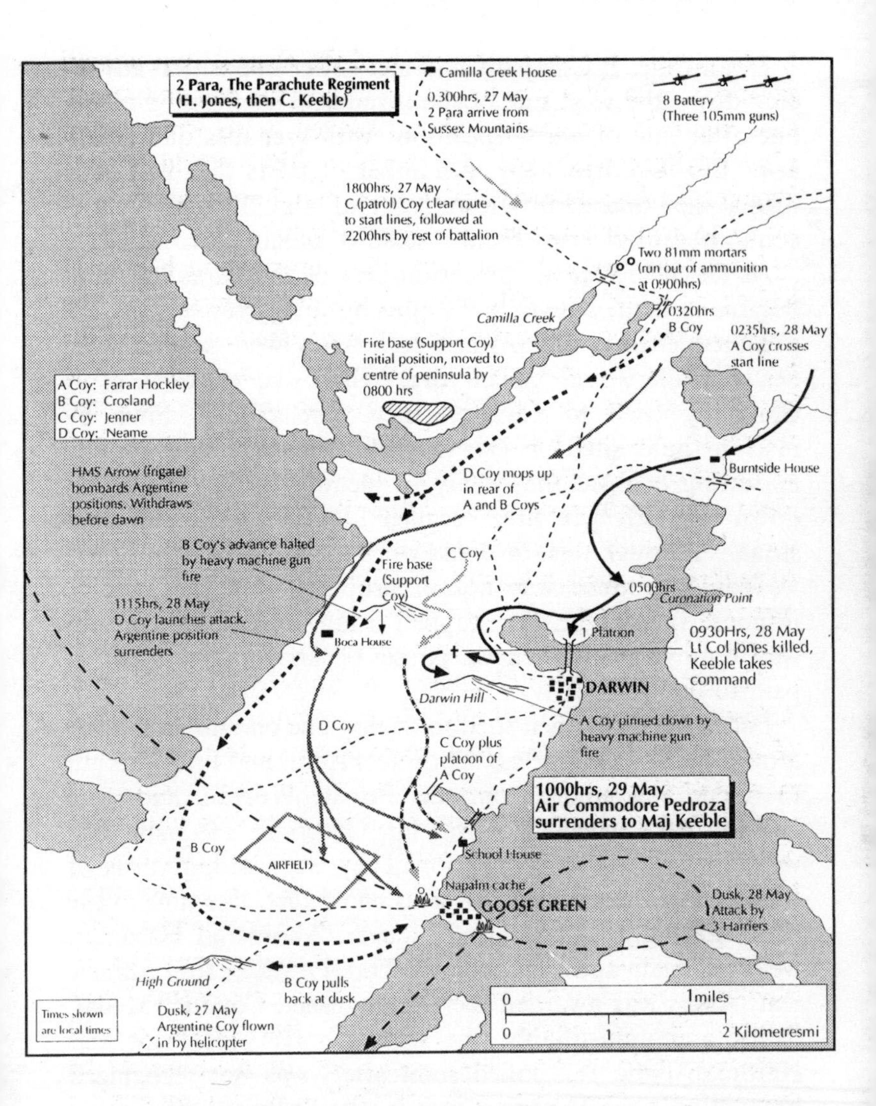

2 Para, The Parachute Regiment (H. Jones, then C. Keeble)
Camilla Creek House
0.300hrs, 27 May
2 Para arrive from Sussex Mountains
8 Battery (Three 105mm guns)
1800hrs, 27 May
C (patrol) Coy clear route to start lines, followed at 2200hrs by rest of battalion
Two 81mm mortars (run out of ammunition at 0900hrs)
Camilla Creek
0320hrs
B Coy
0235hrs, 28 May
A Coy crosses start line
Fire base (Support Coy) initial position, moved to centre of peninsula by 0800 hrs
A Coy: Farrar Hockley
B Coy: Crosland
C Coy: Jenner
D Coy: Neame
Burntside House
HMS Arrow (frigate) bombards Argentine positions. Withdraws before dawn
D Coy mops up in rear of A and B Coys
B Coy's advance halted by heavy machine gun fire
C Coy
Fire base (Support Coy)
0500hrs
Coronation Point
1115hrs, 28 May
D Coy launches attack. Argentine position surrenders
Boca House
1 Platoon
0930Hrs, 28 May
Lt Col Jones killed, Keeble takes command
DARWIN
Darwin Hill
A Coy pinned down by heavy machine gun fire
D Coy
C Coy plus platoon of A Coy
1000hrs, 29 May
Air Commodore Pedroza surrenders to Maj Keeble
B Coy
AIRFIELD
School House
Napalm cache
GOOSE GREEN
Dusk, 28 May
Attack by 3 Harriers
High Ground
B Coy pulls back at dusk
Dusk, 27 May
Argentine Coy flown in by helicopter
Times shown are local times
0
1miles
0
1
2 Kilometresmi

Meanwhile, B Company had met stiffer opposition advancing down the west side of the isthmus. This meant clearing line after line of enemy positions with grenades and small-arms fire. Unfortunately, a number of posts that had been missed then opened fire from their rear, and D Company was committed to clearing them instead of passing through B.

As dawn broke, the precariousness of the Paras' position became evident. Caught in the open by an aggressive and well-prepared enemy, all forward companies were subjected to heavy artillery and mortar fire. When two platoons of A Company tried to advance, they were pinned down by machine-guns sited on Darwin Hill. 'The light now rapidly appearing enabled the enemy to identify targets and bring down very effective fire,' recalled Lieutenant Clive Livingstone. 'Although this too could work for us, the weight of fire we could produce was not in proportion to the massive response it brought. We stopped firing – our main concern was to move away whenever pauses occurred in the attention paid to us.'

Farrar-Hockley then tried to deploy one platoon in a flank attack, but it was hopeless. Corporals Abols and Prior ran out to rescue a wounded comrade, Private Worrall, and were halfway back when Prior was hit. Corporal Hardman then went to help Abols save Prior. They were within yards of safety when Prior was hit by a second bullet, this time in the head. Leaving his body under cover, Abols and Hardman went out again and brought in Worral. For this extraordinary feat, Abols was awarded the Distinguished Conduct Medal. Sadly, and unjustifiably, the selfless bravery of Prior and Hardman (who was killed soon after) was not recognized because only the Victoria Cross is awarded posthumously.

B Company was experiencing similar problems, pinned down by fire from Argentinian positions near Boca House to the north-west of Darwin. 'We were outranged,' said Major

John Crosland, the company commander. 'We just couldn't get across the open ground to get at their machine-guns, and after five hours of fighting, ammunition was critical.'

By now, the two mortars had used all their ammunition while the three 105mm guns were running low on shells. Worse still, the frigate *Arrow* had been forced to return to the relative safety of the San Carlos anchorage, and the Harriers had been unable to take off from their carriers because of heavy fog.

The situation was becoming desperate. Lying under cover next to Farrar-Hockley, Jones realized that Darwin Hill would have to be secured before the advance could continue. 'Dair,' he said, 'you've got to take that ledge.'

Farrar-Hockley at once moved forward with sixteen men, including his second-in-command, Captain Chris Dent. As they moved up the hill they were joined by Jones's adjutant, Captain David Wood. Moments later, met by a storm of fire, first Dent, then Wood and the gallant Corporal Hardman were killed. Farrar-Hockley and the survivors crawled back to cover.

There they met the artillery commander, Major Tony Rice. 'For God's sake come quickly,' he urged. 'The Colonel's gone round the corner on his own.' Armed with a Sterling sub-machine-gun and accompanied by Sergeant Norman and Lance-Corporal Beresford, Jones had charged up a gully towards an enemy machine-gun nest. Within seconds he had been hit in the back of the neck and mortally wounded by a bullet fired from the hill behind him.

It was all so unnecessary because machine-gun and 66mm rocket fire were gradually silencing the enemy positions on Darwin Hill. Before long, white flags appeared and seventy-six Argentinians were taken prisoner, thirty-nine of them wounded.

A few minutes later, in response to an urgent request from 2 Para headquarters to evacuate the stricken Jones, a Scout helicopter was approaching the Darwin area when it was intercepted and blown out of the sky by a Pucara. The Royal Marine pilot, whose brother-in-law was the Marine liaison officer with 2 Para, died instantly. Jones was already dead.

A Boy's Own *Hero*

Lieutenant-Colonel 'H' Jones was one of only two men to win a Victoria Cross during the Falklands War (the other award, to Sergeant Ian McKay of 3 Para, was also posthumous). But it was Jones who received the bulk of the media attention: the *Sun* portraying him as a quintessentially British hero; his lonely charge against the enemy machine-gun in a long tradition of stirring battlefield leadership.

Some army officers were not so easily convinced, arguing with some justification that a battalion commander had no business charging over open ground in such a rash manner. His responsibility was to his men, they pointed out, not to some *Boy's Own Paper* notion of battlefield glory. This was particularly so because his adjutant and right-hand man, Captain Wood, had just been killed (although Jones could not have known this).

Major Keeble, his second-in-command, was more generous. He later explained that Jones had made his move because the battalion had lost its ability to manoeuvre tactically in the face of overwhelming fire. Jones felt it was up to him to break the deadlock. 'He was simply doing what he wanted his battalion to do,' said Keeble.

Sadly, Jones's heroics had no immediate effect on the course of the battle. Stirred by his example, his men did not leap up as one and overwhelm the enemy positions.

This task was accomplished soon after by machine-gun and rocket fire. But if his intention had been to inspire the whole battalion, then his death was not in vain. 'It is impossible,' wrote Max Hastings, 'to believe that, if he were to begin it all again, "H" would not have chosen to do it any other way.'

B company, meanwhile, was still pinned down near Boca House. Even before Jones's death, realizing that a flanking manoeuvre was essential, Major Keeble (who had assumed command when Jones was wounded) had ordered forward the Support Company's Milan anti-tank rocket and sustained-fire machine-gun units to assist. Eventually sited at a range of 1,500 metres, the Milans were fired at Argentinian strong-points with dramatic results. Taking advantage of the enemy confusion, Keeble instructed D Company to advance under the shelter of the shoreline. Then, at 11am, covered by a massive concentration of fire from Support and B Companies, D stormed up the hillside and took the Argentinian position and ninety-seven prisoners.

Keeble now told Farrar-Hockley to advance on Goose Green in support of C Company. Not possible, came the reply. Darwin Hill had to be held in case of a counter-attack, and the severity of A Company's casualties meant he could offer only 3 Platoon. This had to do, and as C Company marched round the hill a patrol was sent into Darwin to winkle out its few remaining defenders with grenades.

But the battle was far from over. As C Company moved down the east side of isthmus, and D and B Companies advanced from the west, the formidable defences around Goose Green and its nearby airfield opened up with everything they had. It was a 'terrifying combination of artillery, mortar, machine-gun and anti-aircraft airburst fire', recalled

Lieutenant Livingstone as he and his men approached the solitary schoolhouse, east of the airfield. 'Little or no cover was available. It was hard to believe that this weight of fire could be maintained for long. It was.'

At one point Major Keeble was caught in the open, snared on a barbed-wire fence, as cannon fire burst around him: 'I remember thinking, I'm losing control. It was not that I was frightened, it was simply that I was the boss, the 2 i/c, trying to maintain the momentum of the attack.'

To break the stalemate, B Company marched round the airfield and moved on Goose Green from the south-west, while C and D Companies made a joint assault on the schoolhouse to the north of the settlement. After a brief firefight, a white flag appeared. Lieutenant Jim Barry of D Company walked forward to take the surrender, but was killed by a burst of gunfire. Enraged, the Paras poured rockets, anti-tank rounds and machine-gun fire into the building, which quickly burst into flames. There were no survivors.

By now the airfield had been taken, its troublesome anti-aircraft guns silenced, and the Paras were closing in for the kill. But as D Company redeployed, it was attacked first by two Skyhawks with bombs, then two Pucaras with napalm and rockets. Fortunately, the explosions and burning jelly were wide of the mark. The two sluggish, piston-engined Pucaras were not so lucky. One was brought down by a Blowpipe hand-held ground-to-air missile, the other by small-arms fire.

Further encouragement was provided by the arrival of three Harriers shortly before dusk. Their cluster bombs devastated the Argentine cannon and gun positions, causing the enemy fire to slacken perceptibly. But as the light faded, help was at hand for the defenders. An Argentinian Chinook and six Huey helicopters landed south of the settlement and dropped off a company of reinforcements. Keeble responded by directing artillery fire on to the area and ordering B Company to move south to block their advance. It was enough to persuade the timorous Argentinians to take cover in the hills. They were picked up over the next few days by patrols from the Paras and the 1st Battalion, 7th Gurkha Rifles.

Yet still the main enemy positions round Goose Green held firm, while the exhausted Paras were low on ammunition and heavily depleted in numbers. To give them some rest and shelter, Keeble ordered his men to pull back off the ridgeline overlooking the settlement. He then radioed Brigadier Thompson and asked for reinforcements and more fire-power. Thompson promised him a company of Royal Marines, three more 105mm guns, the battalion's six unused heavy mortars and plenty of ammunition. In addition, two helicopters were sent to evacuate the casualties under the cover of darkness.

Having been assured by Thompson that the settlement could be destroyed if necessary, Keeble's plan was to call on the enemy to surrender. If they refused he was prepared to unleash a devastating bombardment, using all his considerable firepower, at 9am.

At first light, two prisoners were sent into the Argentinian positions with a letter from Keeble written in Spanish. It informed the Argentine Garrison Commander, Air Vice-Commodore Wilson Dosio Pedroza, that he was surrounded, and that the time had come to release the civilian captives and capitulate. Certain defeat was the only option.

Back came the prisoners with the response that Pedroza had agreed to a meeting. It took place near a little hut on the edge of the airfield between Keeble, accompanied by three officers, an NCO, and two war correspondents as civilian witnesses, and Pedroza and an Argentinian naval officer, as well as the Goose Green Army Commander, Lieutenant-Colonel Italo

Pioggi. Pedroza agreed to release the civilians, but would only surrender after consultation with his superiors and if it was done with honour. Keeble agreed.

Returning to Goose Green, Pedroza signalled his intentions to General Menendez. He was to act as he thought fit, came the reply. So he did just that, marching out at the head of a column of 150 men. In full view of the British, they formed a hollow square, listened to a speech by Pedroza and then sang their national anthem. Finally, they laid down their weapons. But as Keeble walked forward to take Pedroza's pistol, he noticed that all the prisoners were from the Argentinian Air Force. Moments later, an even bigger column, three ranks deep, emerged from the settlement. It was Colonel Pioggi and 900 men. Incredibly, 2 Para had overwhelmed an enemy more than three times its size. In all, 1,200 prisoners were taken and fifty Argentinians killed, with many more wounded. The 2 Para group had suffered seventeen killed (which included Corporal Melia of the Royal Engineers, and Lieutenant Nunn, the Royal Marine helicopter pilot), and thirty-five wounded.

A hasty and unnecessary action against a numerically superior, well dug-in foe, Goose Green serves as a salutary lesson for politicians tempted to interfere in the day-to-day running of military operations. But for the redoubtable fighting qualities of 2 Para, it could have ended in disaster – a military defeat that would have lowered morale at home and encouraged calls for a negotiated peace.

Chapter 4

Misplaced Confidence

'Though thine enemy seem a mouse, yet watch him like a lion,' states the old proverb. In other words, never underestimate even the most primitive and apparently defenceless foe. Sage advice – but too often in the history of warfare it has been ignored, as has the proverb 'He that is secure is not safe.' For a sense of security tends to make an army careless and unaware of potential danger.

So what breeds overconfidence? In most cases it seems to be an innate sense of racial superiority: of the following six examples, five involve white Indo-European armies underestimating their Middle-Eastern, Indian, African and Asian foes. Even the exception, Roman legions fighting German tribesmen in the Teutoburger Wald, conforms to the stereotype in that the vanquished troops regarded their opponents as uncivilized barbarians.

Teutoburger Wald

A false sense of security lay behind the destruction of three Roman legions by Cherusci warriors in the depths of the Teutoburger Wald (forest) in AD 9. It had been five years since the last hint of trouble and Quintilius Varus, the Roman commander, made the mistake of assuming that the Roman Empire's German subjects lacked the moral fibre to rebel. In the meantime, his own troops had grown soft and complacent.

Rome's original conquest of Germany had been an attempt to establish more secure frontiers. Having defeated his rival, Mark Antony, at the naval battle of Actium in 31 BC, Augustus Caesar was the ruler of an empire that stretched from the Atlantic to the Euphrates, from the North Sea to the Sahara Desert. But no border was more fragile that its northern one: running from the mouth of the Rhine to the westward slopes of the Jura Mountains, and then south between the Rhône and the Maritime Alps so that it almost touched the Mediterranean, it could not prevent northern Italy and Macedonia from being raided by the wild tribesmen of the Alps and Danube.

To rectify this, Augustus's legions gradually conquered much of modern-day Austria, Switzerland and Bavaria. By 12 BC, the Rivers Rhine and Danube had together become the Empire's northern frontier. But it was a far from ideal boundary because the two rivers formed a deep salient – like a dagger thrust at the heart of the Empire – with its apex at Basle. This meant that the German tribes within the salient could attack either the Rhenish or Danubian sectors, while the Romans had to defend both, and were forced to move troops around two sides of a triangle if they wanted to reinforce one sector from the other.

Augustus decided to eliminate the salient by pushing the northern frontier 250 miles east to the River Elbe, thereby gaining a relatively straight and easily defended river line from Hamburg to Vienna. But while this would make Transalpine Gaul (the area bounded by the Rhine, the Alps, the Mediterranean, the Pyrenees, and the Atlantic) more secure, there would still be the problem of policing the new buffer zone. Made up of mountains, marshes and endless forests, it was inhabited by fierce semi-nomadic tribesmen described about a century later by the Roman historian Tacitus as large of body, blue-eyed and fair-haired, 'powerful only spasmodically and impatient at the same time of labour and hard work'. Writing even closer in time to the disaster in the Teutoburger Wald, the Greek geographer Strabo noted: 'Against these people mistrust was the surest form of defence; for those who were trusted effected the most mischief.'

The ensuing campaign, conducted by Augustus's stepson Drusus, *legatus* of the Three Gauls (that is, Transalpine Gaul), was a great success. By 9 BC, having cut a swath through the lands of the Sugambri (Lippe valley), Chatti (Nassau), Marcomanni (Thuringia) and Cherusci (Brunswick), his legions had reached the Elbe at Magdeburg. When Drusus was thrown from his horse and killed on the journey home, his elder brother, Tiberius, took over and successfully concluded the campaign in 7 BC.

Bowed they may have been, but the German tribes were far from beaten, and in 1 BC the Cherusci led a revolt. More followed, and by AD 4 the situation in southern Germany was so serious that Tiberius was forced to return to restore imperial authority. Once again the campaign went well, although simultaneous rebellions in Pannonia (Serbia) and Illyricum (Greece) prevented Tiberius from crushing the Marcomanni who had earlier migrated from Thuringia to Bohemia.

When Tiberius left Germany to take on the Marcomanni in 6 AD, Quintilius Varus took command of the Army of the Rhine. A former Governor of Syria, he had married the Emperor's great-niece and owed his appointment to courtly connections rather than military prowess. A contemporary described him as 'a man of mild character and of a quiet disposition, somewhat slow in mind as he was in body, and more accustomed to the leisure of the camp than to actual service in war.'

Although Germany was pacified when he arrived, he made the mistake of assuming that this state of affairs would continue. He was incapable of seeing the difference between the virile Germans and the effete Syrians, and so treated the former with contempt. 'Besides issuing orders to them as if they were actually slaves of the Romans,' wrote the Roman historian Dio Cassius some two centuries later, 'he exacted money as he would from subject nations.' Another critic noted: 'He entertained a notion that the Germans were a people who were men only in limbs and voice, and that they, who could not be subdued by the sword, could be soothed by the law.'

One major source of bad feeling was the Roman practice of levying tribute in gold and silver. For, unlike in Gaul, few luxuries were produced in Germany that could be sold to ensure a flow of specie back into the country. Tribal chieftains were particularly unhappy because precious metals were used mainly to make the ornaments upon which status depended.

To enforce Roman authority, Varus had five legions at his disposal. Two were stationed at Moguntiacum (Mainz) facing the valley of the Main; the other three were near Minden on the Weser during summer; and at Aliso (Haltern) on the upper Lippe in the winter. The summer of AD 9 passed quietly enough, but in September, as Varus was preparing to move to his winter quarters at Aliso, news arrived of a tribal rising. He decided to kill two birds with one stone by making a slight detour to settle the dispute.

Little did he know it, but the rising was part of a clever plan devised by a young Cheruscan, Arminius, to lure the Romans into the depths of the forest where it would be easier to destroy them. The son of Sigimer, Chief of the Cherusci, Arminius had served with a detachment of German troops under Tiberius during the revolt in Pannonia and Illyricum. This had allowed him to observe closely Roman troops in action, and to pinpoint their weaknesses – notably the need for open ground so that they could deploy.

Posted to Varus's headquarters at the age of twenty-six, Arminius despised his Roman overlords and longed to throw off their yolk. He was also anxious to avenge himself on his uncle Segestes, a close supporter of Varus, who had refused him the hand of his daughter Thusnelda (the young couple had been forced to elope). His plot, according to one commentator, was based on the knowledge 'that no one could be more quickly overpowered than the man who feared nothing, and that the most common beginning of disaster was a sense of security'.

Such an elaborate ruse was difficult to keep secret. Segestes learnt about it and told Varus, advising him to put Arminius and the other ringleaders in chains. But Varus declined, seeing the accusation as an underhand attempt by Segestes to settle a family quarrel.

In early October, Varus set off from Minden at the head of the XVIIth, XVIIIth and XIXth Legions, some 20,000 men in all. He was accompanied by Arminius, while the latter's fellow tribesmen acted as an escort. So unconcerned was Varus at the possibility of any trouble that, instead of sending them directly to Aliso, he allowed the soldiers' families to tag along as part of the long baggage train.

The detour to settle the bogus tribal uprising took the Romans directly through the swamps and forests of the Teutoburger Wald, between the Rivers Ems and Weser. Convinced that Arminius would choose this difficult country as the site for an ambush, Segestes again warned Varus; but again his fears were brushed aside. The following day, Arminius and his followers melted away, and reports began to come in that outlying Roman detachments had been slaughtered. Only now would Varus believe the awful truth. He at once ordered the column to make for the road that ran through the Dören Pass to Aliso, where the going would be easier.

But news soon arrived that Aliso was under siege and Varus was forced to make for Münster on the lower Ems. As his troops were hacking a path through the forest, a violent rainstorm added to their woes. According to Dio Cassius, the ground became slippery and 'very treacherous', the 'tops of trees kept breaking off and falling down, causing much confusion'. It was now, with the Romans 'not proceeding in any regular order', but 'mixed in helter-skelter with the wagons and the unarmed', that Arminius chose to attack. All along the length of the huge column, his men flung volleys of javelins into the tightly packed mass.

Nevertheless, the Romans were able to repel the first assaults long enough to enable them to build a rough stockade. Next morning, having burnt most of their wagons, and 'in a little better order', they fought their way into open country. But it was only a temporary respite, for as they continued their march the suffocating forest closed around them once more and the Cherusci continued their attacks. Their losses were terrible.

Again the Romans halted at night. But when they moved off the following morning they were caught by another heavy rainstorm which prevented them 'from going forward and even from standing securely, and moreover deprived them of the use of their weapons. For they could not handle their bows or javelins with any success, nor for that matter, their [leather] shields, which were thoroughly soaked.'

More importantly, the Romans did not have room to deploy into their battle formations and were forced to fight in a disorganized mass. Taking advantage of this, Arminius's warriors moved in for the kill, easily penetrating the Romans' confused ranks. Varus and 'all his more prominent officers, fearing they should either be captured alive or be killed by their bitterest foes', committed suicide by falling on their swords. Only a portion of the cavalry, under their commander Vala Numonius, managed to cut their way out. The rest of the Roman army, the contemporary historian Velleius Paterculus tells us, was 'exterminated almost to a man by the very enemy whom it had always slaughtered like cattle.'

The few who were captured, women and children among them, were either crucified, buried alive or sacrificed to the gods. When Germanicus, Drusus's son, visited the site of destruction a few years later he was met by the gruesome spectacle of whitening bones, fragments of javelins, limbs of horses, and skulls attached to the trunks of trees.

Lucius Caedicius, in command of the camp at Aliso, showed how it could be done by beating off the victorious Cherusci with far fewer men than Varus had had. He later broke out at night and, despite being encumbered by numerous civilians, managed to reach Vetera, where he was met by Lucius Asprenas with two legions. Only now did Arminius and his men withdraw.

As if to confirm the scale of the disaster, the portents of doom were suitably impressive. According to Dio Cassius, the Temple of Mars was struck by lightning, the Alps seemed to fall in on themselves, and 'a statue of Victory that was in the province of Germany and faced the enemy's territory turned about to face Italy.' When Augustus was given the bad news he 'rent his garments . . . and mourned greatly . . . because he expected that the enemy would march against Italy and against Rome itself.' The biographer Suetonius, writing in the early second century, records that 'he let the hair of his head and beard grow for several months', and was much given to banging his head against doorposts, crying, 'Quintilius Varus, give me back my legions!'

With few Roman citizens of military age prepared to enlist, the lost legions were never replaced. Germanicus did manage to reconquer the German tribes and reach the Elbe in AD 13, but it was never a secure border. Before long, the Romans had withdrawn to the Rhine, which once again became the north-eastern boundary of the Empire.

By belittling the German tribes and overestimating the power of his own troops, failing to recognize that they had grown soft through inactivity, Varus allowed himself to be led into country that gave his enemy a great advantage. The resultant disaster dispelled the myth that the legions were invincible, and Roman prestige east of the Rhine received a blow from which it never truly recovered.

The Second Crusade

Unlike its predecessor, the Second Crusade was a fiasco. Launched in the spring of 1147, ostensibly to win back the lost Frankish province of Edessa (in what is now Turkey), it collapsed the following year, having achieved nothing, because its overconfident leaders could not agree on a common objective.

The first call for a Crusade to free Jerusalem, the Holy City, from its Muslim oppressors, the Seljuk Turks, had been made by Pope Urban II some fifty years earlier at the Council of Clermont in November 1095. He was responding to a cry for help from the Byzantine Emperor, Alexius Comnenus, whose subjects had been driven from all but the coastal regions of Asia Minor by the Seljuks, one of several Turkish tribes that hailed from beyond the north-east borders of the Levant.

With cries of 'Deus vult' ('God wills it'), the response was immediate and overwhelming. Commoners and nobles volunteered alike, and all wore a cross on their tunics to denote their holy cause. Among the first to reach Constantinople in 1096 was a rowdy band under the leadership of a popular peacher, Peter the Hermit. Despite Alexius's advice to wait for the main Crusading force, Peter's followers were eager to move on.

Ferried across the Bosphorus on 6 August, they were ambushed at Cibotus soon afterwards and all but annihilated.

The main force of 4,000 mounted knights and 25,000 infantrymen had better fortune. Composed of four major contingents led by princes from the Norman kingdom of southern Italy, and from Toulouse, Flanders, France and Lower Lorraine, they were predominantly French-speaking and became known as the Franks. Before leaving Constantinople, they swore an oath to the Emperor that they would restore any of his conquered territory and swear allegiance to him for any lands occupied beyond the former frontiers.

In July 1097, bolstered by Byzantine troops, the Crusaders defeated the Turks at Dorylaeum, and the following June, after a long siege, they took the city of Antioch. Jerusalem, now held by the Fatimids of Egypt, was invested in June 1099 and successfully stormed a month later. Its first Crusader ruler was Godfrey of Bouillon, Duke of Lower Lorraine. But he died a year later and was replaced by his brother, Baldwin of Boulogne, who was proclaimed King of Jerusalem in November 1100. Over the next eighteen years, Baldwin greatly expanded his domains by conquering the ports of Caesarea, Acre, Beirut and Sidon.

Three other Crusader states had also been created. The nearest to Constantinople was the County of Edessa on the upper Euphrates, populated mainly by Armenians and Syrians. It had been founded by Baldwin and was left in the hands of his cousin, Baldwin of Bourcq, when the former took over in Jerusalem. The Principality of Antioch, the next nearest, had been taken over by Bohemund, a Norman prince from southern Italy, instead of being returned to the Emperor. After his death in 1111, his nephew Tancred assumed power. The last state to be established was the County of Tripoli – sandwiched between Jerusalem and Antioch – by Raymond of St Gilles, Count of Toulouse, in 1109.

But while the Franks controlled the coastal region from Antioch down to the Sinai Desert, it was a thin strip and all attempts to push their borders eastwards – towards the Muslim cities of Aleppo, Homs and Damascus – failed. Instead it was the Sanjuk Turks who made headway when Zengi, ruler of Aleppo, conquered Edessa in 1144.

On hearing the news, Queen Melisende of Jerusalem and Raymond of Antioch at once sent an envoy to Rome to break the news and ask for assistance. In December 1145, Pope Eugenius III responded by issuing a papal bull calling for King Louis VII of France, and all the princes and faithful of that country, to go to the rescue of eastern Christendom. In return, he promised them security for their worldly possessions and remission for their sins.

The pious King Louis, just twenty-six, responded enthusiastically, as did his vassals after hearing Bernard of Clairvaux (who was canonized in 1174, some twenty years after his death) preach a new Crusade at Vézelay in March 1146. Even Conrad III, Emperor of Germany, was eventually won over by Bernard's eloquence and it was his expedition that set off first in May 1147. It numbered around 20,000 armed men and pilgrims, including two vassal-kings, Vladislav of Bohemia and Boleslav IV of Poland, a contingent from Lorraine led by the Bishop of Metz, and many German nobles headed by his nephew and heir, Frederick, Duke of Swabia.

Unfortunately, there was constant friction between the Germans, Slavs and French-speaking Lorrainers, and Conrad, over fifty years old, sick in body and weak in mind, was in no state to control them. Instead, he delegated much of his authority to his ambitious but inexperienced nephew.

On 20 July, having taken an oath of non-injury towards the Byzantine Emperor, Manuel Comnenus, Conrad and his army crossed the Danube and entered imperial territory. But after leaving Sofia, Conrad's men began to pillage the countryside, slaughtering those who stood in their way. Manuel sent troops to police the Crusaders, but this only made matters worse, with Byzantines and Germans frequently at blows.

King Louis, meanwhile, accompanied by his wife, Eleanor of Aquitaine, his brother, the Count of Dreux, and by the Counts of Nevers, Flanders and Toulouse, had set out from Metz in June. More disciplined than the Germans, the French army was just as poorly led, with its young commander easily influenced by his wife and brother. It arrived at Constantinople on 4 October, by which time the unruly Germans had crossed the Bosphorus and were advancing through Anatolia.

Their guide, Stephen, the head of the Emperor's Varangian Guard, advised them to stay within imperial-controlled territory by following the coast road to Attalia. But fatally underestimating their foe, they chose to take the shorter but more dangerous route across the peninsula. On 15 October, Conrad left Nicaea with the main fighting force. For the next eight days, whilst still within Manuel's domains, his army was well fed. As it marched into Turkish territory, however, it lacked food and water.

On 25 October, nearing the River Bathys, close to Dorylaeum, they were surprised by the whole Seljuk army. Many of the German knights had dismounted to rest their horses, and the infantry were tired and thirsty. The battle was no contest, as sudden and repeated attacks by Turkish light horsemen scattered the Germans like sheep. For a time, Conrad tried to rally his men, but by evening he had accepted defeat and fled to Nicaea with a handful of survivors. Nine-tenths of his army and all his supplies were lost. This booty could later be found on sale in the bazaars of the Muslim East.

The French learned of the German defeat when they reached Nicaea at the beginning of November. Having consulted Conrad, who had little option but to join him, Louis chose to advance along the coast road. But this route contained its own miseries. Conrad fell ill at Ephesus and returned to Constantinople. His remaining troops were not so lucky and, with the French, were forced to endure the nightmare of a midwinter march through desolate mountains without adequate supplies. Frequent skirmishes with the Turks were an added hardship. Attalia was reached at the beginning of February, by which time Louis and his knights had had enough. They continued on to Antioch by ship, leaving the increasingly lawless French and German infantry to follow them on foot. Underfed and continually harassed by the Turks, less than half of them made it.

Louis was enthusiastically received at Antioch by Prince Raymond, his wife's uncle, who greatly feared the power of the Turks. Zengi had been murdered by a eunuch of Frankish origin two years earlier, but Nur ed-Din, his son and successor, was equally formidable. Established along the Christian frontier from Edessa to Hama, he had spent the previous autumn stripping the principality of Antioch of all its lands east of the River Orontes. It was no surprise, therefore, when Raymond suggested a joint offensive against Nur ed-Din's power base, the city of Aleppo.

Despite the fact that many of his knights were enthusiastic, Louis hesitated. He was jealous of his wife's obvious affection for her uncle and objected to her passionate support for Raymond's scheme. So, using the excuse of his Crusader vow, which obliged him to go to Jerusalem before undertaking any campaign, he set off towards the Holy City with his reluctant wife in tow. Arriving in May, he found Conrad, now recovered, and his chief princes already in residence.

When all the Crusaders had reached Palestine, Queen Melisende's young son, King Baldwin III, invited them to a great assembly at Acre on 24 June 1148. Notable by their absence were the rulers of the other Frankish states: Prince Raymond was still sulking in Antioch; Count Joscelin of

Edessa was holed up in his remaining possession of Turbessel; Count Raymond of Tripoli had refused to attend after being accused of the murder of the Count of Toulouse, a rival claimant to his lands.

In their absence, the assembly agreed to concentrate all its strength against Damascus. 'It was a decision of utter folly,' wrote a leading historian of the Crusades, Steven Runciman. 'Damascus would indeed be a rich prize, and its possession by the Franks would entirely cut off the Moslems of Egypt and Africa from their co-religionists in northern Syria and the East. But of all the Moslem states the Burid kingdom of Damascus alone was eager to remain in friendship with the Franks; for, like the farther-sighted among the Franks, it recognized its chief foe to be Nur ed-Din.'

To attack Damascus was, as Runciman noted, 'the surest way to throw its rulers into Nur ed-Din's hands'. But the barons of Jerusalem were greedy for the fertile lands of the Burid kingdom, and were still smarting from their failed invasion of Damascene territory the year before. To the Crusaders, Damascus was a city hallowed in Holy Writ; Aleppo, on the other hand, held no spiritual significance.

Despite the losses suffered in Anatolia, and the lack of assistance from the other Frankish states, the Christian army that set out from Galilee in mid-July was still the greatest such force ever: about 50,000 strong. It set up camp on the edge of the orchards and gardens that surrounded Damascus on the 24th, by which time Unur, the Burid ruler, had ordered his provincial governor to send reinforcements and had dispatched a messenger to Aleppo to ask Nur ed-Din for urgent assistance.

Next day, the Damascene army attempted to hold the Crusaders at the village of al-Mizza, to the south of the city, but superior numbers soon forced them to retire behind the city walls. By afternoon, the Crusaders were in possession of all the orchards and had begun to cut down trees to make palisades. Before long, with a rejuvenated Conrad in the van, they had forced their way to Rabwa, on the River Barada, right beneath the walls of the city. Convinced that the end was near, the citizens of Damascus began to barricade the streets for the final struggle.

But help was on its way. Next day, the reinforcements summoned from the provinces began to pour into the city through its northern gates. At once, Unur launched a counter-attack, pushing the Crusaders back from the walls. The assaults continued during the next two days, while guerrillas infiltrated the Christian forces through the orchards. With their camp now under constant attack, the Crusader leaders decided to move their army to the plain beyond the east wall where the Turks would find no such cover. It was a mistake, for the new site lacked water and faced the strongest section of the city wall.

With his troops increasing in number all the time, and Nur ed-Din on his way, Unur stepped up his attacks. It was now the Crusader camp, not Damascus, that was under siege. Yet still the Christian leaders found time to quarrel over the future of Damascus once it was captured: Queen Melisende and the barons of Jerusalem wanted it to be incorporated as a fief of the Kingdom, with Guy Brisebarre, Lord of Beirut, as its ruler; Thierry of Flanders, on the other hand, aspired to hold it as a semi-independent fief, like Tripoli, and he had the support of Conrad and Louis, and also of King Baldwin, whose half-sister was Thierry's wife.

When the local barons discovered that the kings favoured Thierry, they lost interest in the enterprise. It was later rumoured that they had been in secret contact with Unur, and that vast sums of money, found to be counterfeit, were passing between Damascus and the Court of Jerusalem. Unur may well have told them that if they retired at once he would cancel his alliance with Nur ed-Din. For Nur ed-Din was already at Homs, negotiating the terms of his aid. One of his main conditions was that his troops be admitted within the city walls, something which Unur was understandably anxious to avoid. But in any case, the Franks had little option. They could expect no reinforcements, whereas Nur ed-Din's arrival was imminent.

On hearing the latter arguments, Conrad and Louis were shocked and accused the local barons of disloyalty. Without the barons' assistance, however, they could do nothing and so, reluctantly, they ordered the retreat. At dawn on 28 July, just four days after their arrival, the Crusaders struck camp and retired towards Galilee. But if Unur really had bought the Christians off, he did not let them retire in peace, using light horsemen armed with bows to harry them every inch of the way. By the time the once-proud army reached the safety of Palestine, in early August, thousands of dead men and horses littered the road to Damascus.

The fact that such a great host had abandoned its objective after just four days of fighting was a major blow to Christian prestige; the aura of invincibility that had been built up by the glorious First Crusade was shattered for ever. Yet the Second Crusade need not have ended as it did. 'No medieval enterprise started with more splendid hopes,' wrote Runciman. 'Planned by the Pope, preached and inspired by the golden eloquence of St Bernard, and led by the two chief potentates of western Europe, it had promised so much for the glory and salvation of Christendom.'

Instead, it only managed 'to embitter the relations between the western Christians and the Byzantines almost to breaking-point, to sow suspicions between the newly come Crusaders and the Franks resident in the East, to separate the western Frankish princes from each other, to draw the Moslems closer together, and to do deadly damage to the reputation of the

Franks for military prowess.' And all because of the 'ineffectual folly' of its leaders, who could not agree on a common purpose and who made the cardinal error of underestimating a redoubtable foe.

Custer's Last Stand

The slaughter of General Custer and much of his US 7th Cavalry by Sioux and Cheyenne Indians near the Little Bighorn River, Montana, on 25 June 1876, has long been portrayed in American history as an heroic defeat. In fact, it was a squalid episode – from which only the Indians emerge with any credit – characterized by Custer's naked ambition, lack of regard for his men, and foolish contempt for his foe.

George Armstrong Custer was born in New Rumley, Ohio, in December 1839, the son of a local farmer and blacksmith. But his grandfather, a Hessian officer named Küster, had fought for the British in the War of Independence, and from early childhood Custer's wish was to be a soldier. Thanks to the patronage of the local Congressman, this wish was granted in 1857 when he was sent to the US Military Academy at West Point. His time there, however, was less than distinguished. In four years he accumulated an incredible 726 demerits – known as 'skins' – and graduated 34th and last in the class of 1861. West Point had never been afflicted with a less promising pupil, commented one biographer, and Custer himself remarked that his time there should be studied by future cadets as an example to be avoided.

Things soon went from bad to worse. A few days after graduating in June, when on duty as officer of the guard, he failed to stop a fist fight between two cadets. Found guilty by a court martial of dereliction of duty, his career might have been over but for the fact that the Civil War was in progress.

Ordered to report for duty with the 2nd Cavalry (in the Northern army), he was cited for bravery during the Union defeat at the First Battle of Bull Run in July 1861.

War was the perfect backdrop for Custer's own brand of reckless gallantry, and it secured him rapid promotion. In 1862, during the Peninsula Campaign, he waded across the Chichahominy River in full view of Confederate troops in order to see if it was fordable. Hearing of this 'act of desperate gallantry', the Union commander, General McClellan, appointed him to his staff with the rank of captain of volunteers. In June the following year, after more derring-do at the Battle of Aldie, he was appointed brigadier-general of volunteer cavalry.

More battles came and went, and with each Custer's fame grew. In October 1864, aged just twenty-four but already the most trusted lieutenant of the famous cavalry leader, General 'Little Phil' Sheridan, he was promoted to the temporary rank of major-general of volunteers and given command of the 3rd Cavalry Division. He used it, the following spring, to dog the retreat of Lee's army from Richmond, capturing prisoners, wagons and guns in the process. And it was to Custer that the Confederate flag of truce was conveyed at Appomattox on the morning of 9 April 1865. 'I know of no one,' wrote Sheridan, 'whose efforts have contributed more to this happy result than those of Custer.'

He had become a national hero, instantly recognizable by his long reddish-gold curls and non-regulation uniform – often a costume of olive-gray velveteen, lavishly decorated with gold braid, a cavalier hat and a bright red neckerchief. This latter item he always wore so that his men could recognize him in the heat of battle. It was often to their cost, as the eighty-six casualties of a wild charge against an entire Confederate division would testify.

With the war over, however, Custer's star began to wane.

Demobbed from the volunteer service in February 1866, he lost his honorary rank and resumed the substantive rank of mere captain. In July, he was promoted, for the last time, to lieutenant-colonel and given active command of the newly formed 7th Cavalry (the actual commander, a colonel, never got round to leading it in the field). Officially, therefore, he remained a lieutenant-colonel, although it was customary to address an officer by the highest rank he had achieved.

In 1867, after taking part in the ineffective campaign under General Winfield Scott Hancock to subdue the marauding Cheyennes, he was again court-martialled on charges of absence without leave (while visiting his sick wife) and the unauthorized execution of deserters. Surprisingly found guilty, given the mitigating circumstances and lack of evidence, he was sentenced to be suspended for one year. But General Sheridan, his old mentor, intervened and he was recalled early to join a new expedition against Black Kettle's Cheyennes. On 27 November 1868, Custer led his men in a surprise dawn attack against the Cheyenne chief's village near the River Washita. It was largely unopposed, although Custer later claimed that 103 braves had been killed. In fact, only eleven were fighting men, including Black Kettle himself; the rest were squaws, children and old men, slaughtered like cattle. This was Custer's sole 'victory' on the Plains.

It was also in 1868 that the Federal government signed a treaty with the Sioux, reserving to them for all time a huge tract of land comprising the western half of what is now South Dakota. But all this changed six years later when an expedition led by Custer found gold in the Black Hills – part of Sioux territory. This led to an inevitable influx of covetous settlers, which in turn caused the Sioux and their Cheyenne allies to hunt and raid outside their own lands.

By late 1875, large numbers of Sioux had left their reservation and moved east into the wild tract of land between the

Yellowstone River and the Bighorn Mountains in south Montana. There, under Chief Sitting Bull, a Hunkpapa Sioux, they joined forces with the Cheyenne. President Ulysses S. Grant reacted by authorizing the War Department to undertake punitive action. The final plan, drawn up by General Sheridan, was for three columns to converge on the area in which Sitting Bull's braves were operating. General George Crook, with 800 men, would move north from Fort Fetterman in eastern Wyoming; Colonel Gibbon, with 450, would move south-east from Fort Shaw in west Montana; and General Terry, in overall command with 1,250 men, would move west from Fort Abraham Lincoln in Dakota Territory.

Custer had expected to command the third column. But in mid-March 1876, summoned to Washington to testify before a Congressional committee investigating fraud in the Indian Service, he made comments highly critical of William Belknap, the former Secretary of War. President Grant, a personal friend of Belknap, responded by forbidding Custer to join the campaign. It was only through the intercession of Generals Terry and Sheridan, who valued Custer's presence, that Grant relented and allowed him to lead his regiment under Terry's command.

The first column to see action was Crook's. On 17 June, three weeks after leaving Fort Fetterman, it ran into a part of the renegade band, numbering about 1,000 Cheyenne and Sioux warriors near to the Rosebud Creek in southern Montana. Although the Indians had the element of surprise, having made a fast night march from their camp to the north, Crook's men managed to hold their ground and kill about thirty-five braves; American dead were nine troopers and one Shoshone scout. But the following day, fearing another attack, Crook decided to withdraw, and the Indians therefore considered the encounter a victory.

Terry's column, meanwhile, had left Fort Abraham Lincoln

on 17 May. It comprised the entire 7th Cavalry under Custer, three and a half infantry companies, a battery of Gatling guns, a wagon train and a beef herd. In all, there were 1,018 soldiers, 30 Ree scouts and 190 civilians employed by the Army. The Indian camp was said to be near the Little Bighorn River, which flowed north into the Bighorn, and thence into the Yellowstone River. Terry decided, therefore, to march west until he reached the Yellowstone, and then follow the river upstream.

By mid-June, Terry had met up with Gibbon's column, and together they continued on up the Yellowstone River. A few days later, a scouting party under Major Reno, Custer's second-in-command, discovered an Indian trail leading from the Rosebud Creek in the direction of the Little Bighorn River. At about the same time, Terry's scouts had seen smoke rising from the Little Bighorn valley. It seemed as if they had located the hostiles' camp.

That night, Terry, Gibbon and Custer met aboard the steamer, *Far West*, moored at the confluence of the Yellowstone and Rosebud Creek, and decided on a plan of attack. Custer's 7th Cavalry, the strongest and most flexible unit, would ride south along the Rosebud to a position east of the suspected Indian camp. Terry and Gibbon, with the rest of the men, would follow the Yellowstone to the Bighorn, and the latter to the Little Bighorn. The Indians would be caught in a trap. The tentative date for it to be sprung was 26 June.

On hearing the details of the plan from General Terry, Major Brisbin of the 2nd Cavalry did not think that Custer's force of 675 men was strong enough. He offered to send four of his own companies with the mobile column, and begged Terry to go in command. But Terry declined, saying that Custer had been hurt by Grant's censure and needed a chance to vindicate himself with the 7th Cavalry alone. When Brisbin persisted, Terry told him to speak to Custer and offer his whole regiment. If he agreed, Terry would take command. So

Brisbin did just that, warning Custer that scouts had estimated a total Indian force of at least 3,000 braves. Custer's predictable response was that the 7th could handle anything (he later told his officers that he expected to meet no more than 1,500 Indians). As a last resort, Brisbin suggested taking some Gatling guns. Custer agreed, but an hour later had changed his mind, saying that the gun carriages would slow his march.

Brisbin was not the only concerned officer. Gibbon, who had been an artillery instructor at West Point when Custer was a cadet, feared that the cavalryman's 'zeal would carry him forward too rapidly'. Shortly before Custer set off on the 22nd, he said to him: 'Now, Custer, don't be greedy, but wait for us.'

His exact response is unknown. All Gibbon would say is that, 'He replied gaily as, with a wave of his hand, he dashed off.'

Custer's orders were specific. 'Instead of proceeding at once into the valley of the Little Big Horn, even should the trail lead there,' recalled Gibbon, Custer was told to 'continue on up the Rosebud, get closer to the mountains, and then striking west, come down the valley of the Little Big Horn.' Needless to say, he ignored them.

On 24 June, while still moving down the Rosebud, Custer's scouts picked up an Indian trail which they followed to the site of a recent camp. The marks of numerous lodge circles indicated a force far larger than Custer had been led to expect. Further up the creek, the trail was more than a mile broad, the ground so furrowed by travois poles (a travois is a kind of horse-drawn shed) that it looked like a ploughed field. The scalps and beards of white men were found by Crow scouts and taken to Custer. He was visibly moved, and signalled to the scouts that he had been sent by the Great Father in Washington to conquer these Indians who were killing white men. He might be killed, he indicated, but the Sioux would feel his wrath.

They camped for the last time that afternoon. Custer was

anxious to get to grips with the enemy and had ordered a night march up the long slope towards the ridge separating the Rosebud from the Little Bighorn. It did not seem to bother him that he was still 20 miles short of the point where he was supposed to cross over. Terry later stated that had Custer lived he would have been court-martialled for disobedience.

At 11pm, with Ree and Crow scouts leading the way, they set off up the slope in darkness, making little attempt to conceal the noise of horses' hooves, rattling equipment and braying pack mules. Early next morning – Sunday, 25 June – the scouts located a knoll that overlooked the valley of the Little Bighorn. One was sent back to inform Custer who, with the rest of the regiment, had stopped 10 miles to the east and was breakfasting on coffee and bacon. Smoke from these campfires could be seen from the knoll, which meant it could also be seen by any Indians in the vicinity. Clearly, Custer was unconcerned that the enemy might know of his advance.

Arriving with the regiment at the knoll in mid-morning, their coming heralded by a rising dust cloud, Custer climbed part of the way up and borrowed an old telescope belonging to one of the scouts. Though he was unable to see the Indian village, his eagle-eyed scouts assured him that the valley was so covered with teepees that it seemed to be draped in a white sheet. Mitch Bouyer, a half-breed scout, told him that it was the largest gathering of Indians he had seen in more than thirty years. Custer, however, seemed unconcerned and ordered the regiment to advance; they would attack once the exact location of the Indians had been determined.

About noon, having just passed over the ridge, Bouyer advised Custer to be cautious; they would find more Indians than they could handle. Custer told him that if he was afraid he could stay behind. Bouyer's reply was that he would go wherever Custer went, but that if they did enter the valley they would not come out alive.

In the early afternoon, still on the downslope some twelve miles south-east of Sitting Bull's village, Custer ordered a halt. Despite still being unaware of the exact size and location of the enemy camp, and knowing that Crook and Gibbon would not be in a position to assist him for at least another day, he was determined to attack. As if that was not bad enough, he now made the even more fatal decision to divide his command and advance from different directions. Captain Benteen would take three companies and search the badlands (that is, a deeply eroded barren area) to their left; if he came across any Indians he was to pitch into them and let Custer know. Major Reno would take another three up the valley before crossing the Little Bighorn; he, too, was to attack if he met Indians. Custer would follow Reno with five companies, and support any assault. Captain McDougall would stay behind with one company to protect the mule train.

Benteen was unconvinced. 'Hadn't we better keep the regiment together, General?' he asked. 'If this is as big a camp as they say, we'll need every man we have.'

'You have your orders,' Custer replied.

Benteen later testified that he did not believe Custer had any definite battle plan, and that his own 'senseless' task was to go 'valley hunting ad infinitum'. After riding up and down hills for ten or fifteen miles, with the horses exhausted, he decided that the Indians were too wise to struggle over such rough ground and decided to turn round. With the order, 'Right oblique!' Benteen angled his men back towards the valley.

Reno, meanwhile, riding well ahead of Custer, had crossed the Little Bighorn and spotted the huge Sioux-Cheyenne encampment spread out across the valley floor. Following Custer's instructions, he continued to advance with the intention of attacking its southern end. But with 500 yards still to go, hundreds of mounted Sioux emerged from the village, causing him to halt, dismount his men and form a

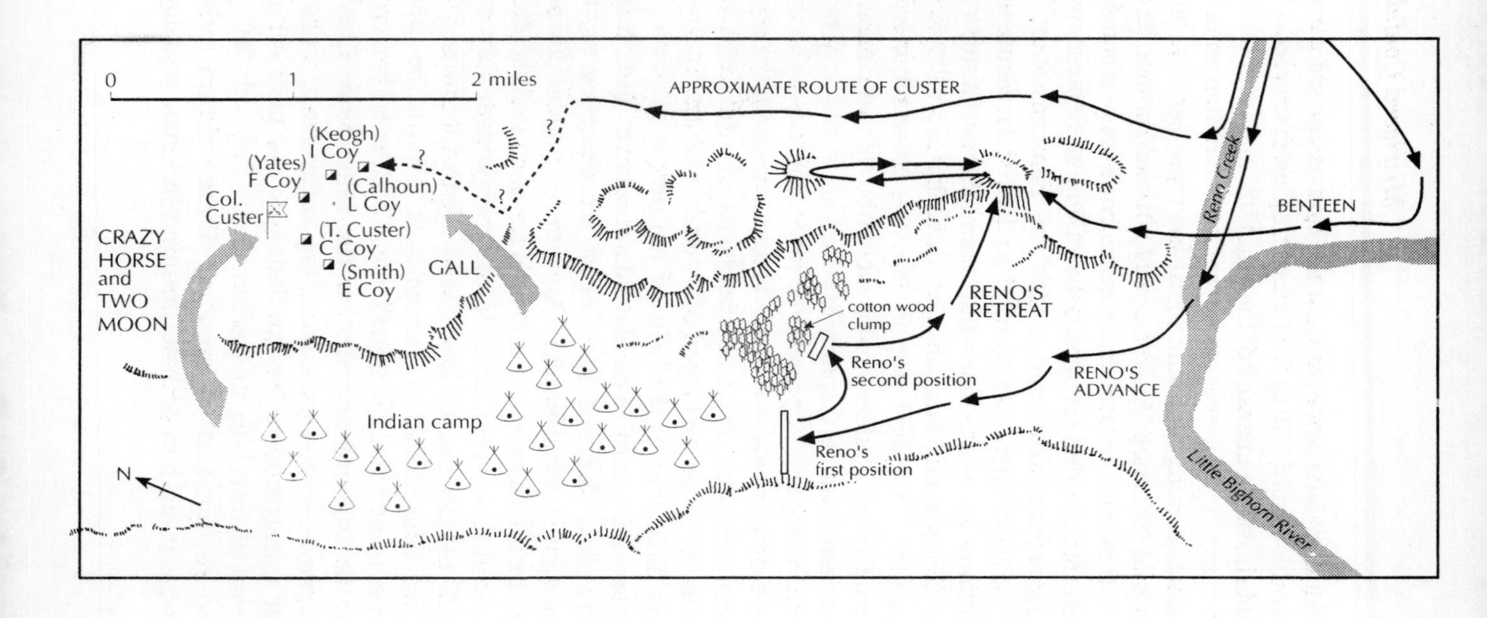
0
1
2 miles
APPROXIMATE ROUTE OF CUSTER
(Keogh)
I Coy
(Yates)
F Coy
(Calhoun)
L Coy
Col.
Custer
(T. Custer)
C Coy
(Smith)
E Coy
CRAZY
HORSE
and
TWO
MOON
GALL
BENTEEN
Reno Creek
RENO'S
RETREAT
cotton wood
clump
Reno's
second position
RENO'S
ADVANCE
Indian camp
Reno's
first position
N
Little Bighorn River

skirmish line. Two troopers were unable to stop and careered on into the village, where they were dragged from the saddle, stabbed, beaten and scalped.

Their comrades faced a similar fate. Outnumbered and outgunned – many of the Indians had repeating Winchester rifles while the troopers were armed with single-shot Springfields – they were in danger of being outflanked when Reno ordered them to remount and withdraw to a nearby clump of cottonwood trees in a bend of the Little Bighorn River. Once there, he asked Custer's favourite scout, Bloody Knife, what the Indians would do next. But before Bloody Knife could answer, a bullet hit him in the head, spattering his brains over Reno's face. The shock of this is said to have panicked the American officer, who shouted out: 'Any of you men who wish to make your escape, follow me!'

So began the wild retreat out of the woods, down the valley, over a fordable point of the river and up on to the relative safety of a nearby bluff. Pursued all the way by yelling Indians, many troopers were dragged off their horses and butchered. Others, those who never received the order to retreat, were killed in the woods. Once the soldiers had fled, squaws, children and old men rushed out of the village to strip and mutilate the bodies. Black Elk, a thirteen-year-old Oglala Sioux, scalped one soldier while he was still alive. But his knife was blunt and the short-haired soldier made such a fuss, grinding his teeth, that Black Elk had to shoot him before he had finished.

The majority of Reno's command eventually ascended the bluff, although eleven of them were wounded. They left behind thirty-two dead. Reno was one of the first to reach safety, but he had lost his distinctive straw hat and is said to have thrown his revolver away once it was empty.

On hearing the sound of firing, Custer did not ride immediately to Reno's assistance. Instead, he sent a messenger to hurry along the mule train with the extra ammunition, before turning right off the trail into the hills bounding the east of the village, in an attempt to attack it from the flank. Galloping north for about a mile and a half, Custer halted his men under the shadow of large hill. Then he, three officers and his orderly joined the scouts on the crest of the ridge from where they could observe the village through field glasses.

The massive camp, made up of six circles, each containing a separate self-governing tribe, seemed to be asleep. But for a few dogs and ponies, and a handful of squaws and children, it looked deserted, denuded of fighting men. 'Hurrah, boys, we've got them!' exclaimed Custer. 'We've caught them napping. We'll go down and make a crossing and capture the village.'

What Custer did not realize was that the lodges were full of warriors who had stayed up half the night celebrating the victory over Crook. If he had, he might not have waved his hat so gaily as he galloped back to his troops. On they rode in a northerly direction, the intervening hills screening them from the village. A mile further on, Custer stopped and turned to his orderly, Private Martini. 'I want you to take a message to Captain Benteen,' he said. 'Ride as fast as you can and tell him to hurry. Tell him it's a big village, and I want him to be quick, and to bring the ammunition packs.'

Adjutant Cooke, not trusting the Italian immigrant's English, scribbled a quick message. It read: 'Come on. Big village. Be quick. Bring packs.'

The last white man to see Custer alive and live to tell of it, Martini met Benteen on his return from the badlands. But as they hurried towards the battleground, Benteen noticed Reno's command organizing a defensive position on the bluff and he decided to join them instead. Technically, he disobeyed orders, but whether his three companies could have saved Custer is doubtful. It is more likely that they would have met the same fate, as would Reno's men if they had not been reinforced. Benteen later excused his inactivity by saying that any attempt to assist the general would have been 'suicide'.

Contrary to popular image, Custer did not cut a particularly dashing figure on the day of his death. It was hot and he had taken off his buckskin jacket and rolled it up behind his saddle. He was wearing, instead, a not particularly fetching navy blue army-issue shirt, his trademark red neckerchief and buckskin trousers tucked into his boots. His famous curly locks had recently been cropped and could hardly be seen beneath his broad cream-coloured hat, its right brim fastened to the crown to enable him to sight a rifle. He wore two revolvers but no sword (nor did his men, as all swords had been left behind with the heavy baggage).

Thus attired, Custer led his men down the hill with the intention of crossing the river opposite the centre of the huge Indian camp. But as they neared the Little Bighorn, they were fired on by a mass of charging Hunkpapa Sioux under Chief Gall and forced up on to the ridge to their right. Minutes later, Crazy Horse's Oglala Sioux and Two Moon's Cheyennes appeared from the head of the village and blocked their line of retreat.

Only L Company, led by Custer's brother-in-law, the handsome Lieutenant James Calhoun, seemed to put up any organized resistance. It was the first to be attacked by the swirling mass of Indians, and after the battle next to each corpse lay a pile of twenty or thirty spent shells. Elsewhere, remarkably few were expended. After their horses had been stampeded by Indians shouting and waving blankets, many of Lieutenant Smith's E Company tried to escape by running down a narrow gully. It was a dead end and twenty-nine bodies were discovered there, shot and clubbed to death.

Most of the 7th Cavalry died in five disorderly company clusters, roughly in the shape of a V – an arrowhead – with

Custer at the point. But towards the end, company structure disintegrated as men fought in smaller and smaller units, and bodies were scattered along the ridge for hundreds of yards. Custer died with about forty men and half of the officers around him, indicating that this was the last stand. Several had turned their horses loose; others had shot them to form barricades, and a number of the dead animals were found in a fairly symmetrical ring about fifteen yards in diameter.

According to the Cheyenne, the last living bluecoat was a big officer with a curly moustache. The battle was over and squaws, boys and old men were busy stripping and mutilating corpses when he raised himself on his left elbow and began to wave a pistol in his right. A Sioux warrior calmly walked over, took the pistol from his feeble grip and shot him through the head.

Custer was hit by two bullets: in the left side and the left temple. Other than that his naked body was unmarked and had not even been scalped. Rain in the Face, a Sioux brave (one of the nine said to have been Custer's killers), is supposed to have recognized him and prevented his mutilation. Most of his men were not so lucky. A few feet away his brother Tom, the commander of C Company, lay face down, bristling with arrows, the back of his head smashed. He had been scalped, his throat cut, and his abdomen split both horizontally and vertically so that the entrails emerged. Mark Kellogg, a correspondent from the Bismarck *Tribune*, was found in long grass minus his scalp and one ear; Lieutenant Cooke, the Adjutant, had one of his bushy sideburns cut off by Wooden Leg, a Cheyenne, who then tied it to an arrow shaft and presented it to his delighted grandmother.

Who killed General Custer?

The exact circumstances of Custer's death at the Little Bighorn have never been determined. There were no survivors to tell the tale, while Indian accounts are wildly contradictory; no less than nine braves are mentioned in tribal folklore as having killed 'Long Hair'. One of the more credible accounts of his last moments was given by an Arapaho named Waterman who rode with the Cheyenne. As he reached the battle he saw Custer at the top of the ridge on his hands and knees. 'He had been shot through the side and there was blood coming from his mouth. He seemed to be watching the Indians moving around him. Four soldiers were sitting up around him . . .'

Lieutenant James Bradley, commanding General Terry's Crow scouts, was the first white man to discover Custer's corpse. 'His expression was rather that of a man who had fallen asleep and enjoyed peaceful dreams,' he wrote, 'than of one who had met his death amid such fearful scenes as that field had witnessed, the features being wholly without ghastliness or any impress of fear, horror or despair.'

He had been shot twice: in the left side beneath the heart and in the left temple. The side wound was bloody and had probably killed him; the hole in his temple was clean, almost certainly a *coup de grâce*. Captain Benteen, who had a good look at his body, was of the opinion that the first bullet was fired at some distance from a rifle, either a Henry or a Winchester.

Sergeant Knipe said he found him lying across two or three dead soldiers with only part of his back touching the ground. He was naked but for his socks, as were the

cont. overleaf

men around him. The bottom of one of his boots lay nearby; the top had probably been cut off by a squaw to make moccasins.

He did not appear to have been mutilated. At least that is what the public were told. Unpublished letters, however, gave details of various disfigurements: slashed thighs, slit ears and arrows fired into the groin. This information was withheld out of respect for his wife Elizabeth.

Why he had not been scalped is a mystery. One version is that his killer, a Sioux brave called Rain in the Face, insisted that his body be left untouched in recognition of his courage. A more likely explanation is that his sparse hair, prematurely balding and cut short, was not considered worth keeping.

Of the 225 officers and men who rode with Custer down to the village, not one survived. Curly, a Crow scout, claimed to have done so, but in reality he and the other Indian scouts peeled away before the final advance. Mitch Bouyer is alleged to have told him to 'go to the other soldiers [meaning Terry's men] and tell them that all are killed. That man [pointing to Custer] will stop at nothing. He is going to take us right into the village.'

Reno and Benteen did eventually ride out with their men to see what had become of Custer; but on witnessing the tail end of the massacre they hurried back to the bluff, pursued by Indians, who trapped them there. The siege continued for most of the next day, by which time Benteen had effectively taken over command from the indecisive Reno. Sensibly, the Indians did not try to rush the position, preferring to pick off the bluecoats with accurate rifle fire; in all, twenty-four were killed and forty-eight wounded. The torment finally ended during the afternoon of the 26th, when the Indians dismantled their huge village and headed south. The following morning, the troops under Terry and Gibbon arrived.

In all, the 7th Cavalry lost 281 killed and 59 wounded at the Battle of the Little Bighorn. The Indians, in comparison, suffered only 40 fatalities, although many more braves would have died of their wounds. Even so, the disparity in casualties between soldiers and untrained hostiles is startling. No greater proof of Custer's folly, his arrogant underestimation of the Indian's fighting capacity, and his overweening ambition to win a famous victory before the main force caught up with him, is needed.

Isandhlwana

At Isandhlwana, on 22 January 1879, armed only with spears, the Zulus inflicted on the British the most catastrophic reverse ever suffered by a modern army at the hands of men without guns. Victorian England was aghast. How could it happen? The answer was simple: far from being untrained savages, the Zulus were disciplined and organized – indeed their whole society was geared to war – and Lord Chelmsford, the British commander, had made the cardinal error of not according their capabilities proper respect.

In the 1820s, thanks to the military genius of King Shaka, the Zulu nation had become the most powerful in southern Africa, with an empire encompassing much of modern-day Natal. Twelve years of bloody conquest left millions dead and thousands of square miles uninhabited. Into this vacuum moved white settlers.

By the time King Cetshwayo, Shaka's nephew, succeeded to the throne in 1873, Zululand had been hemmed in on two sides by the rapid expansion of these European communities. To the south lay the British colony of Natal; to the east the Boer

Republic of the Transvaal. But the British were the biggest threat because they aspired to unite the whole of South Africa under a single authority: their own. The policy was known as 'Confederation' and Sir Bartle Frere, appointed Governor of the Cape and first High Commissioner for South Africa in 1877, was given the task of implementing it.

Shortly after his arrival in Cape Town in April 1877, he received word that Britain had annexed the Transvaal, declaring it a Crown Colony. His main priority now was to nullify the military threat posed by the Zulu kingdom, and he could only do this by a war of conquest. But the Tory government of Benjamin Disraeli, Earl of Beaconsfield, embroiled in yet another Balkan crisis and a war in Afghanistan, had little enthusiasm for a fresh conflict. Treat the Zulus with 'a spirit of forbearance', Frere was told. But he had no intention of drawing back and, using slow communications as an excuse, hoped to be able to present London with a *fait accompli.*

The excuse he needed was provided by two minor border incidents. In July 1878, two wives of Sihayo, a prominent Zulu induna (chief), fled to Natal with their lovers. In two separate incidents, Sihayo's brother and three sons pursued the women, dragged them back and executed them. The lovers were unharmed. On a subsequent occasion a Colonial Service surveyor, Smith, and a white trader, Deighton, were apprehended and held by Zulus for an hour on a small island near the Zulu bank of the River Tugela.

Frere had already discussed the desirability of war with the commander of Imperial troops in South Africa, Lieutenant-General the Hon. Sir Frederic Thesiger (he became Lord Chelmsford on the death of his father in October). On 11 November, the day he was given permission to raise 7,000 native troops by the reluctant Sir Henry Bulwer, Lieutenant-Governor of Natal, Chelmsford told Frere that he was now ready to invade Zululand 'should such a measure become necessary'. It was the last piece in Frere's jigsaw.

Exactly a month later, Frere's representatives met Cetshwayo's at the Tugela's Lower Drift. The former's purpose was to present the findings of a Boundary Commission which had been arbitrating on rival Boer and Zulu claims to the Blood River territory, on Zululand's eastern border. The Commission had actually found in favour of the Zulus, recognizing their sovereignty but giving the Boers residential rights, but Frere had made the findings conditional on the acceptance of a separate ultimatum, the terms of which were to be met in thirty days. Typically rambling in style, it boiled down to thirteen demands.

The first three, to be met within twenty days, called for Sihayo's brother and three sons to be handed over for trial in Natal; 500 head of cattle to be paid in compensation; and a further fine of 100 cattle for the offence against Smith and Deighton. The next ten were far more severe and struck at the very independence of the Zulu nation. They included disbanding the Zulu army; abandoning Shaka's military system; outlawing summary execution; and accepting a permanent British resident at Ulundi, the Zulu capital. Frere knew that Cetshwayo could never agree to such severe terms.

Fearful of their reception, Cetshwayo's representatives took two weeks to reach the royal kraal (Afrikaans for a native village) at Ulundi. But news of the ultimatum arrived long before them thanks to John Dunn, a European who lived in Zululand and who was regarded as a friend by the Zulus. Cetshwayo and his Great Council dismissed the more draconian demands, preferring to placate Britain by complying with the first three. On 18 December, a message was sent via Dunn: Cetshwayo agreed to surrender the men and pay the cattle fines but asked for more than twenty days because the rivers were in flood and it would take time to collect the cattle from Sihayo.

Frere denounced the response as a 'pitiful evasion' and insisted that his troops would enter Zululand if the first three demands were not met on time. Cetshwayo, of course, was unable to keep to the deadline and on 4 January 1879, Frere placed the further enforcement of all demands in the hands of Lord Chelmsford. A week later, with the lapse of the thirty-day ultimatum, Frere announced that British troops 'are crossing into Zululand to exact from Cetywayo [*sic*] reparation for violations of British territory', and 'to enforce compliance with the promises, made by Cetywayo at his coronation, for the better government of his people'.

Lord Chelmsford, the man with the task of conquering Zululand, was the 51-year-old son of a former Lord Chancellor. Since joining the 95th Foot (the Rifle Brigade) at the age of seventeen, he had served in the Crimean War, the Indian Mutiny, the Abyssinian campaign, and more recently had brought the Ninth Kaffir War to a successful conclusion. Even so, his career was described by one biographer as 'light on action and heavy with peacetime staff duties'. Tall and handsome, the possessor of a fine spade beard, he was regarded by his superiors as competent and reliable, but not particularly talented. To his inferiors, he was stiff, reserved and uncommunicative.

Nevertheless, with an army of almost 17,000 men, even a general as plodding as Chelmsford could feel reasonably confident. His main problem was how to protect Natal and the Transvaal while at the same time deploying enough force to defeat the Zulus. The prolonged campaign against the Kaffirs had taught him that the first blow in native warfare must be a heavy one. (Now a derisive term, originally from Arabic, for a black African, 'kaffir' was formerly used for the members of any of a number of Bantu-speaking tribes of southern Africa.) Yet a single massive column advancing into Zululand would not encourage the Zulus to attack; they might, instead, invade a defenceless Natal. He therefore decided to march on Ulundi with three columns. One would cross at the Lower Drift of the Tugela and head north; a second, accompanied by Lord Chelmsford himself, would move north-east from Rorke's Drift on the Buffalo River; the third would march south-east from the headwaters of the Blood River. Two further columns would be kept in reserve: one at the Middle Drift of the Tugela below Krantz Kop, to guard the road to Pietermaritzburg; and one at Luneburg in the Transvaal, to prevent a Zulu invasion and keep an eye on the sullen Boers.

Frere had deliberately timed his ultimatum so that Chelmsford's invasion would coincide with the Zulu harvest, when many warriors would be busy collecting grain. But on this occasion the rains had been late and the harvest was not ready. The First Fruits ceremony, however, had gone ahead as normal, meaning that the Zulu regiments were fully mobilized at Ulundi before the invasions began.

During the morning of 11 January, Chelmsford's centre column was ferried across Rorke's Drift into Zululand. It was made up of two battalions of imperial infantry (less one company to guard the drift), a battery of Royal Artillery, a company of Royal Engineers, two battalions of the Natal Native Contingent (NNC), a company of the Natal Native Pioneer Corps, and a hotchpotch of white volunteer mounted troops. In all, there were 1,275 imperial infantrymen, 320 irregular cavalry, 132 Royal Artillerymen to service six 7-pounders, and 2,566 Natal Kaffirs commanded by white officers and NCOs. Unfortunately, none of the natives had uniforms and only one in ten was armed with a rifle – the rest had spears, knobkerries (clubs) and shields. Including staff and civilian drivers, the total was 4,709 men. To drag the guns, transport the supplies, or provide food were more than 1,500 oxen, 50 horses, 67 mules, 220 wagons and 82 carts.

Next day, Chelmsford sent a strong force to the nearby Bashee valley to raid the stronghold of Sihayo, the Zulu induna whose relatives had sparked the war. Sihayo and most of his men were at Ulundi, but a small force had been left behind. Its stern resistance was finally overcome when British troops scaled the flanking cliffs. Having burnt Sihayo's homestead and rounded up his cattle, the raiding party returned to their camp on the Zulu bank of the Buffalo.

The plan was for Chelmsford to establish his first base at Isipezi Hill, 20 miles east of Rorke's Drift. But it would take time because the intervening terrain was both rocky and marshy, and the engineers had to make the track passable for transport wagons. The best place for an interim camp, his scouts reported, was at the foot of a hill known to the Zulus as Isandhlwana – about 10 miles from the drift.

As the track climbed out of the wet lowlands of the Bashee valley, it crested a broad saddle between a stony koppie (small hill) on the right and Isandhlwana on the left. It then wound its way east towards Isipezi Hill (part of the Nkandhla Range) through a rectangular plain, 4 miles wide and 8 long. To the south of the plain lay the Malakata and Inhlazatye Hills, with broken country beyond them that led down to the valley of the Buffalo; to the north was a steep escarpment, rising several hundred feet to a plateau that was bounded by the Nquthu Range. This plateau could be reached by several steep ravines, the largest of which began just beyond a large conical koppie that rose from the plain 1½ miles in front of Isandhlwana.

The hill itself lay north to south, with a broad spur stretching from its low northern end up to the plateau, 1,500 yards away. Between Isandhlwana and the conical koppie, the open ground fell gently away to a wide, shallow donga that cut across the plain. It was here, in the shadow of the mountain, that Chelmsford set up camp on 20 January.

Major Lonsdale, Commandant of the 3rd NNC, placed his

two battalions furthest to the left of the track, level with the northern end of the hill. Next came the 2nd Battalion, 24th Foot (2/24th), the Royal Artillery, the mounted units and finally, to the right of the track, on the slopes of the stony koppie, the 1st Battalion, 24th Foot (1/24th). Chelmsford placed his staff tents in the centre, behind the 2/24th, while the main wagon park was situated astride the track on the saddle itself. The whole camp stretched for 800 yards across the eastern slopes of Isandhlwana and the saddle.

Before the campaign had begun, Chelmsford had issued a number of field regulations. The first of these ordered that every halting place should be turned into a defensible position, either by digging entrenchments or by laagering (drawing into a defensive circle) the wagons. Yet when Lieutenant-Colonel Glyn of the 2/24th Foot suggested protecting the camp at Isandhlwana, Chelmsford demurred. It was only a temporary halt, he said, the ground was too stony to dig and he needed the wagons free to keep up the flow of supplies from Rorke's Drift. In any case, he did not fear an attack on a camp housing almost two imperial battalions. As long as they had time to form, each company in two ranks, the front kneeling and the rear standing, their stinging volleys would provide all the firepower necessary.

To provide ample warning of an attack, Chelmsford had pushed out a number of infantry pickets. The furthest were more than a mile away, beyond the stony koppie and up the spur to the lip of the Nquthu plateau. Though contracted at night, the cordon still ringed Isandhlwana and the stony koppie, and a large detachment of NNC remained on the spur.

Chelmsford had also ordered up reinforcements from the column at the Middle Drift. Commanded by Colonel Anthony William Durnford, Royal Engineers, Durnford, Natal's Chief Engineer, they included five troops of the Natal Native Horse (NNH), three companies of NNC and a rocket battery. They reached Rorke's Drift on the 20th.

That same day, Chelmsford led a small scouting party towards the Mangeni gorge, south-west of Isipezi Hill, but encountered no Zulus. Early next morning, to make doubly sure that the ground to his right was clear, he sent Major Dartnell with 200 of his volunteer cavalry to search the Nkandhla Hills, and Commandant Lonsdale with six companies of the NNC to scout the Malakatas.

In the afternoon of 21 January, Dartnell bumped into several hundred Zulus in the hills before Isipezi, where the plain joined the far end of the Nquthu escarpment. After retiring back across the Mangeni, he despatched three officers to Chelmsford to ask for reinforcements and permission to attack in the morning. He also sent a message to Lonsdale to join him, despite the fact that neither force had made provision for a night in the open.

Shortly before dusk, a party arrived from Isandhlwana with blankets and rations and a note from Chelmsford giving Dartnell permission to attack at his discretion. As the light began to fail, Dartnell could see groups of Zulus slip over the crest of the next ridge, 3 miles to the east, and settle down for the night. A mounted patrol was sent into the valley. Narrowly avoiding encirclement, it reported that the Zulus numbered about 1,500 warriors. This was too big for Dartnell's force to handle, so he sent back a second message asking Chelmsford to reinforce him with several companies of imperial infantry.

Overtaken by darkness, the messenger did not reach the camp until 1.30am. Chelmsford was immediately woken up. As he himself had scouted the Nquthu plateau for a distance of 5 miles the previous day, seeing only sixteen Zulu riders, he concluded from Dartnell's message that the main impi (force, especially one ready for battle) was at the head of the plain. With only 200 mounted men and Lonsdale's 1,400 Natal Kaffirs, Dartnell was too weak to oppose it. He decided, therefore, to split the column and march at once to Dartnell's assistance. It was a fatal error, based on inadequate scouting and a conviction that the force left behind would be strong enough to defeat any Zulu attack.

At 3.30am on the 22nd, Chelmsford departed the camp with six companies of the 2/24th, four artillery pieces and the NNC Pioneer Corps company. Lieutenant-Colonel Pulleine of the 1/24th was left in command with five of his companies, one of the 2/24th, 600 Natal Kaffirs, 100 mounted troops and two 7-pounders. Chelmsford had also sent an urgent message for Durnford to move up to the camp from Rorke's Drift with his 250 mounted natives, 300 native infantry and the rocket battery. This would give Pulleine a total force of over 1,700 men, including 600 imperial infantrymen. His defensive arrangements included vedettes (mounted sentries) about a mile from the camp on the Nquthu plateau and the conical koppie to the east, and a long curve of infantry pickets a short way behind.

Shortly after 8am, as the camp was settling down to breakfast, one of the vedettes rode in with the news that a large force of Zulus was advancing from the north-east across the plateau. The 'Fall in' was sounded and the infantry pickets were brought in (except for a company of NNC on the lip of the plateau commanding the Isandhlwana spur). Within ten minutes the men were drawn up in two lines in front of the tents, with the flanks refused (withdrawn or moved back from regular alignment with the main body) to cover the Malakatas and the spur.

Pulleine also sent a hastily scribbled note to Chelmsford: 'Report just come in that the Zulus are advancing in force from left front of the camp. 8.5 A.M.'

Handed the note at around 10am, Chelmsford, by then some miles away, decided to do nothing. He considered Pulleine's force strong enough to look after itself, and did not think that he could return in time to make a difference.

Meanwhile, at 9am, Pulleine had received another message from the vedettes to the effect that the Zulus had been in three columns; two had retired out of sight to the north, the other was moving north-west. Shortly afterwards, Durnford arrived. Superior in rank to Pulleine, he was now in nominal command. As the last report had indicated that the Zulus were retiring, he ordered the men to finish their breakfast.

Soon after, the vedette on the Nquthu plateau reported that the Zulus were retiring along the plateau to the east. By this time, Durnford had already sent two troops of Natal Native Horse up on to the plateau to search for the three Zulu columns. One, led by Lieutenant Roberts, moved north-west, the other, under Lieutenant Raw and accompanied by Captain Shepstone, Durnford's political agent, went east. But no sooner had these horsemen disappeared from sight than firing broke out from the direction of the vedette.

Durnford now came to the extraordinary conclusion that while the main impi was still in the vicinity of Isipezi Hill, there was now a large force on the plateau retreating to the east, and consequently threatening the left rear of Chelmsford's column. He therefore decided to take the balance of his Native Horse, a company of NNC and the rocket battery across the plain in an attempt to prevent these Zulus from rejoining the main impi. If Raw and Shepstone managed to drive them over the edge of the escarpment, he would catch them in the open.

Before leaving, Durnford asked for two companies of the 1/24th to accompany him. Pulleine demurred, feeling that his superior had already overstepped Chelmsford's orders, and Durnford did not insist. Pulleine did agree, however, to reinforce the NNC picket at the head of the spur with a company of the 1/24th under Lieutenant Cavaye.

Durnford left at 11am, having sent instructions for Shepstone to press the attack. When the message arrived, Shepstone sent Raw to search the dongas and shallow ravines to the north-east; Roberts, who had found nothing to the left, was sent along the rim of the plateau.

Eventually, a group of Raw's men spotted a few Zulus herding cattle up a slope, a full 4 miles from Cavaye's men at the head of the spur. Giving chase, they lost sight of the Zulus who ran over the crest of the slope. As the foremost rider reached the top of the rise, he had to rein in sharply to prevent falling over the edge of a deep ravine. Down below, sitting in complete silence, covering every inch of the sides and floor of the wide Ngwebeni valley, were 20,000 warriors. The main impi had been located.

It had left Ulundi on the 17th with orders to march slowly, so as not to tire itself, and to attack only in daylight. Upon no pretext was it to invade Natal. On the 20th, it had camped to the north of Isipezi Hill, moving into the Ngwebeni valley the following day. Its intention had been to stay there during the 22nd; a new moon was due and to attack on the day of a 'dead' moon was inauspicious. The warriors were hungry and detachments had been sent out to scour the small kraals on the plateau for cattle and grain. These were the groups that had been spotted by Pulleine's vedettes.

Now the whole impi had been discovered and, as one, the warriors rose to their feet. First over the lip was the umCijo regiment, closely followed by part of the uThulwana and by Qetuka, second-in-command of the Undi corps. Shouting commands, Qetuka sent both regiments after the retreating horsemen.

Next out to the right of the umCijo were the uNokenke and the regiments of the uNodwegu corps: the umKhulutshane, the uDududu and the IsaNgqu. They spread out and made for the head of the spur. To the left, the uVe, the inGobamakhosi and the uMbonambi raced forward on the flank of the umCijo.

The impi commanders, Mavumengwana and Tswingwayo, tried to prevent the stampede so that they could organize the squatting semi-circle from which the classic 'chest-and-horns' formation would evolve (the chest to engage the enemy frontally, the horns to race round on either flank to encircle them) – but it was too late. They were only able to intercept the uDloko and the rest of the Undi corps under Cetshwayo's brother, Dubalamanzi, which had been resting in another valley a little further back. These men they formed into the 'loins', or tactical reserve, which followed the chest.

Raw's troop fell back in the face of the advancing horde, stretching for more than a mile from wing to wing, firing as they went. They met Shepstone, who sent a message to warn Durnford before riding back to the camp via the spur. He blurted out the news to Pulleine, who had just received a message from Chelmsford to strike the tents and send on the baggage to join him. Seemingly unable to appreciate the danger he was in, Pulleine sent back a note to Chelmsford that there was 'heavy firing' to the left of the camp and that he was unable to move 'at present'.

Pulleine's best option now would have been to concentrate his troops in a solid formation behind the wagon and horse lines, with the steep flank of the mountain at their backs. But instead of withdrawing Cavaye's exposed company at the head of the spur, he sent another, Lieutenant Mostyn's, to support him. He also failed to order the opening of the heavy boxes containing the reserve ammunition.

By the time Mostyn's men reached the top of the spur, the Zulus' right horn was pouring across Cavaye's front at a distance of 600 yards. Undeterred by the volleys directed at them, they were heading towards the rear of Isandhlwana. The vedette and the two troops of Natal Native Horse had by now reached the spur and they, too, joined in the shooting.

Shepstone's messenger reached Durnford as he was advancing across the plain. No sooner had he halted than the Zulus' left horn appeared on the edge of the plateau. For a distance of almost two miles, stretching back towards the spur and far beyond his own position, the heights were black with Zulus. As he watched, they poured down a number of ravines on to the plain. To his right, the inGobamakhosi and uMbonambi raced towards the camp, followed by the uVe. To his left the umCijo and umHlanga charged round both flanks of the koppie, and before long had overrun the rocket battery which was lagging behind. Its supporting company of NNC fled back to the camp.

Durnford began to retire, his men firing into the inGobamakhosi as the left horn of the Zulus tried to outflank him. On reaching the wide donga, a mile from the camp, he decided to make a stand. He was soon joined by about thirty men from the Natal Carbineers, and so accurate was the fire from this position that the Zulu's left horn was stopped in its tracks.

Pulleine, meanwhile, had sent the two 7-pounders to engage the Zulus on the plateau from a rocky knoll about 600 yards in front of the camp. To support the guns, Lieutenant Porteus's company of the 1/24th was placed on their left, Captain Wardell's on their right. Beyond Wardell was the sole company of the 2/24th, under Lieutenant Pope, and then a mixed force of the NNC. Pulleine had also withdrawn the two companies from the spur, and Cavaye was now protecting the north of the camp, some way to Porteus's left. Two hundred yards in front of the wide gap between these two companies, and inadvertently forming the knuckle of the defensive line, were 300 men of the NNC who had been on outpost duty.

Extending the line to the left of Cavaye was Mostyn, the contingent of Native Horse, the NNC picket and, finally, Lieutenant Younghusband's company of the 1/24th below the spur. The remaining NNC men were formed in a loose body directly in front of the camp. Firing was general all along the line and the Zulu advance seemed to have been halted.

Durnford, in the donga way out to the right, was in the most danger and Pulleine now ordered Pope to give him long-range supporting fire. But Durnford had been in action for some time and his men were beginning to run out of ammunition. His runners could not find their own supply wagons and the infantry quartermasters refused to issue ammunition to natives. The infantry companies were also running out of ammunition, and it did not help when the quartermaster of the 2/24th, responsible for just Pope's company, refused to issue cartridges to runners from the other companies. Instead, he sent them 500 yards further to where the 1/24th quartermaster was issuing ammunition.

As Durnford's rate of fire fell off, the inGobamakhosi and uVe moved further to their left and began to edge forward. In danger of being outflanked, and virtually out of ammunition, Durnford ordered his men to mount up and retire, joining the rest of the mounted troops in the mouth of the saddle. It was the turning point of the battle.

No longer faced by Durnford's murderous fire, the inGobamakhosi and the uMbonambi rose and moved forward. Pope's company, its flank now exposed, wheeled to face them. But all along the line the fire was slackening, and as one the Zulus leapt up and charged, rattling their assegais against their shields and shouting 'uSuthu!' It was too much for the NNC on the knuckle, who threw down their weapons and fled to the rear, taking the NNC in front of the camp with them. They were preceded by the two 7-pounders, which had been limbered up and sent to the rear.

Through the gap left by the Natal Kaffirs poured the umCijo, followed by the isaNgqu and the umHlanga. First Cavaye's company, then Mostyn's, was taken in the rear. There were no survivors. Younghusband's men made a fighting withdrawal along the back of the camp and up the slopes of the mountain. Sixty of them reached a rocky ledge where they fought to the last man.

The remaining companies along the front of the camp were doomed. Forced back by the charging Zulus, their ammunition exhausted, they rallied in little groups, fighting with rifle butt and bayonet until they were overwhelmed. As the Zulus began to pour into the camp, slaughtering every man they came across, Pulleine entered the guard tent of the 1/24th, grabbed the battalion's Queen's Colour and handed it to Lieutenant Melville, the adjutant, telling him to save it at all costs. He then re-entered the tent, sat down at his desk and began to write a letter. He was still writing when a Zulu burst in. Picking up a revolver, he shot the warrior in the neck, but the wounded Zulu still managed to lunge over the desk and kill Pulleine with an assegai thrust.

In all, about 400 men, mainly Natal Kaffirs, made it over the saddle while two groups of men, led by Shepstone and Durnford, held the Zulu horns apart. But eventually these gallant pockets ran out of ammunition and they were killed to a man in desperate hand-to-hand fighting. The remnants of Pope's company, the last of the imperial infantry, died a short distance away.

'I have eaten!'

The Zulu warrior's main weapon at Isandhlwana was the assegai (spear). It came in two forms: the light throwing assegai, 6 feet long, with an effective range of up to 70 yards; and the short stabbing assegai with its heavy, broad blade. The latter weapon had been developed by King Shaka for close-quarter fighting and was known as the *iKlwa* (the sucking sound it made as it was withdrawn). As the victim slid off his blade, the warrior would shout 'Ngadla!' (I have eaten!). In addition to his *iKlwa* and a couple of throwing assegais, he might

cont. overleaf

also carry a knobkerrie, a spindly stick with a heavy burled end that was used as a club.

For defence the warrior had an oval cowhide shield, 3 feet wide and long enough to cover his entire body. Light and tough, it could deflect airborne spears, turn the point of a thrusting assegai or even be used to batter an opponent into submission. By hooking its left edge over the left edge of an opponent's shield, a manoeuvre introduced by Shaka, its bearer could spin his enemy to the right with a powerful backhand sweep. He would then sink his assegai into his foe's exposed left armpit.

British firepower at Isandhlwana was mainly provided by the imperial infantry, armed with the single-shot, breech-loading Martini-Henry rifle. Firing a heavy .455 calibre soft lead bullet that could drop an elephant, it was a devastating weapon in trained hands, accurate to more than 1,000 yards. Volley fire against massed targets would often open at 800 yards, and even an average marskman could score hits at half that distance. But the rifle had a vicious kick, especially when the barrel was fouled, and tended to bruise shoulders and cause nosebleeds. Also, during prolonged firing in which the weapon became very hot, the thin brass cartridge had a habit of jamming in the breech and had to be eased out with a knife.

The only other weapon possessed by the ordinary soldier was a triangular bayonet known as the 'lunger'. Officers carried swords and centre-fire, double-action .455 revolvers, usually made by Adams or Webley.

Those who had escaped the slaughter were horrified to find warriors of the Undi corps – the tactical reserve – astride the track to Rorke's Drift. To avoid them, they veered to the left, plunging down the broken ground that led to the Buffalo River. They were pursued by Zulus, who even crossed to the far bank in their effort to leave no survivors. It was here that Lieutenant Melville, accompanied by Lieutenant Coghill, having lost the Queen's Colour in the water, was finally killed, as was Coghill. The Colour was recovered miles downstream sometime later, and thus never fell into enemy hands.

Of the 1,700 men in the camp, only 55 whites and around 300 Natal Kaffirs survived. All 21 officers and 581 men of the 24th Foot were killed. In accordance with Zulu custom, the corpses were stripped and disembowelled, to release spirits which might otherwise haunt their killers. The Zulus had lost about 1,500 killed, although many more would later die of their wounds.

The casualties were the worst ever suffered by a British army fighting a native foe, and they were largely due to Chelmsford's underestimation of his enemy. Nothing else can explain his failure to fortify the camp at Isandhlwana, to scout the Nquthu plateau on the day of the attack, and to return to Pulleine's assistance when he was informed that a Zulu force was advancing.

Yalu River

On 25 November 1950, as American-led United Nations (UN) troops and their South Korean allies advanced towards the Yalu River, the border between North Korea and China, they were attacked by more than 180,000 Chinese soldiers. Within days the UN forces were in headlong retreat and a defensive line was only re-established deep inside South Korea. And all because US General Douglas MacArthur, the UN commander, and his political masters had ignored the warning signs and dismissed the Chinese military threat as negligible.

The seeds of the Korean conflict were sown during the last

weeks of the Second World War when, at the Potsdam Conference in July 1945, the US President, Harry S. Truman, gave Stalin a free hand to deal with Japanese troops still in Korea and Manchuria. America was preoccupied with the difficulties of invading mainland Japan and, in any case, saw Korea as of no long-term strategic interest.

All this changed when the dropping of two atomic bombs on the cities of Hiroshima and Nagasaki, on 6 and 9 August respectively, brought about the unconditional surrender of Japan. US troops were now available to take part in the occupation of Korea – and their presence would help to prevent the whole country from falling under Communist influence. So Truman's government suggested to the Soviets that they limit their advance to the 38th Parallel, a horizontal line roughly halfway down the country, chosen by two Pentagon officers on the grounds that below it lay the capital, Seoul, more than half of the population, the best agricultural land, and most of the light industry. To Washington's amazement, the Soviets agreed.

So came into being Soviet-sponsored North Korea and US-influenced South Korea. But ultimate reunification was always the intention, and in December 1945 the Russians accepted an American proposal to make Korea the object of a Four-Power 'International Trusteeship' for five years, paving the way for independence as a unified state.

Left to its own devices, however, it became increasingly obvious that Korea would succumb to the Communists. In September 1947, therefore, in the face of Soviet objections, the US referred the future of Korea to the United Nations. With the Eastern Bloc abstaining, the General Assembly voted unanimously for a UN supervision of elections to a Korean government, followed by independence and the withdrawal of all foreign troops.

With the Russians and North Koreans against UN involvement from the start, and the left-wing parties in South Korea refusing to take part in the elections, the result was a farce. Dr Syngman Rhee's 'Association of the Rapid Realization of Independence' gained most seats (partly thanks to widespread intimidation), and in July 1948 he became the South Korean Republic's first elected leader. Over the coming months his ruthless regime would more resemble a dictatorship than a democracy.

In June 1949, despite Rhee's protestations, the last American troops left South Korea. The Soviets had already pulled out of North Korea, leaving behind a rigidly disciplined Stalinist government under their protégé, Kim Il Sung. In September 1948, the Democratic People's Republic of Korea was proclaimed.

Relations between the two embryo countries gradually worsened with a series of border incidents. Finally, on 25 June 1950, sanctioned by Russia and with the connivance of Communist China, North Korea invaded the south. The People's Army was composed of 135,000 well-motivated troops, supported by an armoured brigade of Russian T34 tanks and plenty of artillery. Its air force had just 200 Russian Yak-9 fighters and Il–10 ground-attack bombers, but these were more than a match for Seoul's tiny supply of T-6 trainers which were destroyed in the first hours of the war.

The corrupt Republic of Korea (ROK) army was just 95,000 strong, with no tanks, anti-tank weapons, or artillery heavier than 105mm. A third of its vehicles were in disrepair, and its ammunition reserve was for just six days.

The result was inevitable. Within hours of crossing the 38th Parallel, ten communist divisions had brushed the feeble ROK opposition aside and were driving south in four armour-tipped columns. That evening, the UN Security Council – minus the Soviet delegate, who had walked out the previous January in protest at the UN's refusal to seat Communist China in place of the Chinese Nationalists who, by then, occupied only Formosa (now Taiwan) – passed unanimously a resolution condemning the attack and calling for the withdrawal of North Korean forces south of the 38th Parallel.

Two days later, the Security Council passed a US-sponsored resolution, calling upon member nations to 'render such assistance to the Republic of Korea as may be necessary to repel the armed attack and to restore international peace and security to the area'. Only Communist Yugoslavia abstained (the Soviet delegate still staying away).

Without even bothering to ask Congress to declare war, President Truman ordered his troops to assist South Korea as part of a UN 'police action'. General MacArthur, Supreme Commander of the Allied Powers in the Pacific, based in Tokyo, was put in charge and told to conduct a limited war. Communist aggression was to be repelled at all costs, but the battlefield was to be confined to the Korean peninsula. But MacArthur, an ambitious man who had made his name commanding US forces in the Pacific during the Second World War, and who had taken the Japanese surrender, had other ideas. He did not believe in limited war, regarding the fight against communism as one to the death, and thought that the general in the field should be given complete autonomy over the deployment and use of his forces.

His first act was to rush three US divisions – designated the Eighth Army – to South Korea from the Army of Occupation in Japan. Unfortunately, thanks to heavy post-war cuts, these divisions were understrength, undertrained and under-equipped, and could do little to stem the North Korean advance.

By the end of July, the Eighth Army survivors and the remnants of the ROK Army had been driven back into a small bridgehead in the south-east corner of the peninsula known as the Pusan Perimeter. Thanks to reinforcements, including the

US Marine Brigade and the British 27 Brigade, the perimeter held – just.

Then, on 15 September, MacArthur launched his masterstroke: an amphibious landing by the 70,000 men of X Corps at Inchon, halfway up the western coast near Seoul. It took the North Koreans completely by surprise, and as X Corps drove east, the Eighth Army began its long-awaited break-out from the Pusan Perimeter. Attacked from two directions, the North Korean front collapsed as troops began to melt away, discarding weapons and equipment as they went.

On 26 September, American troops linked up at Osan. The following day, Seoul, the capital of South Korea, was recaptured. Already, 125,000 prisoners had been taken and the People's Army was in full retreat. It was a miracle. Only a month earlier, the Eighth Army had been facing total defeat; now, thanks to MacArthur's genius, the boot was on the other foot.

The issue now facing the US government, its allies and military commanders, was how to gain the greatest political advantage from their success. The obvious next step was to cross the 38th Parallel, destroy the remnants of the North Korean Army, and unite Korea under a 'democratic' government. But the UN mandate had only extended to the ejection of communist troops from South Korea, and this had now been achieved. Furthermore, the Soviet delegate had resumed his seat at the Security Council and would undoubtedly veto any extension of that mandate.

Without referring the matter to the Security Council, the US government authorized MacArthur to finish off the People's Army in North Korea. There were, however, a number of crucial provisos: he was not to go ahead with operations if major Soviet or Chinese forces (or both) proved a threat; his forces were not to cross the Manchurian or USSR borders of Korea (since 1948, Manchuria had been part of Communist

China); no non-Korean ground forces were to be used in the north-east provinces bordering the Soviet Union or in the area along the Manchurian border; and, finally, he was not to support his operations by using air or naval power against Manchuria or against Soviet territory.

On 9 October, the Eighth Army crossed the 38th Parallel. For a week, the North Korean troops put up a stern resistance. Then they broke, fleeing north, with the US 1st Cavalry and 24th Infantry Divisions in hot pursuit. With total victory seemingly in his grasp, MacArthur was called to a meeting with President Truman on Wake Island in the Pacific.

He assured the President that formal resistance in North Korea would end 'by Thanksgiving' (November) and that he would withdraw the Eighth Army to Japan by Christmas, leaving only X Corps as an occupation force. In the unlikely event of the Chinese intervening, his air force would commit 'the greatest slaughter'. If the Russians supported the Chinese with planes, their incompetence would cause them to 'bomb the Chinese as often as they would bomb us.'

In other words, Truman need have no worries about either the Chinese or Russians getting involved. MacArthur's assurances seem to have had the desired effect. 'I've never had a more satisfactory conference since I've been President,' Truman told reporters.

On 19 October, Pyongyang, the capital of North Korea, fell to the 1st ROK Division. Kim Il Sung and his Communist government had already fled. The UN troops, meanwhile, were continuing their largely unopposed advance. On 25 October, X Corps made a second amphibious landing at Wonsan on the east coast of North Korea. There were no casualties because ROK troops had taken the city two weeks earlier (the day before the US 1st Marine Division splashed ashore, Bob Hope had given a show intended for US troops).

But as the Eighth Army moved ever closer to the Chinese border, and the British proposed the establishment of a buffer zone south of the Yalu River, jointly policed by the Chinese and the UN, MacArthur made clear his contempt for diplomatic niceties. 'The widely reported British desire to appease the Chinese communists by giving them a strip of North Korea,' he wrote to his superiors, 'finds its historic precedent in the action taken at Munich on 29 September 1938.'

On 24 October, MacArthur issued a new directive removing any remaining restrictions on the deployment of American troops towards the Manchurian border. His senior commanders were 'authorized to use any and all ground forces as necessary, to secure all of North Korea'. When the Joint Chiefs of Staff in Washington queried this order, he ignored them. When they requested that he issue a statement guaranteeing the safety of China's vital Suiho electricity-generating plant in North Korea, he refused, saying he could not be sure that the plant was not powering communist arms production.

That MacArthur was not immediately sacked was because he was a talisman. Quite apart from the fact that he was a hero of the Second World War, he, and he alone, had been responsible for the 'miracle of Inchon'. Outright military victory was now tantalizingly close and American politicians were intoxicated by the scent of success. In any case, MacArthur had assured them that he could easily deal with any Chinese offensive. 'At the root of American action,' wrote Max Hastings in *The Korean War*, 'lay a contempt, conscious or unconscious, for the capabilities of Mao Tse-tung's nation and armed forces.'

The Americans certainly had plenty of warning. On 2 October, the Chinese Premier, Chou En-lai, had informed the Indian Ambassador that China would intervene if the United Nations crossed the 38th Parallel. Truman dismissed the threat as 'a bald attempt to blackmail the UN'.

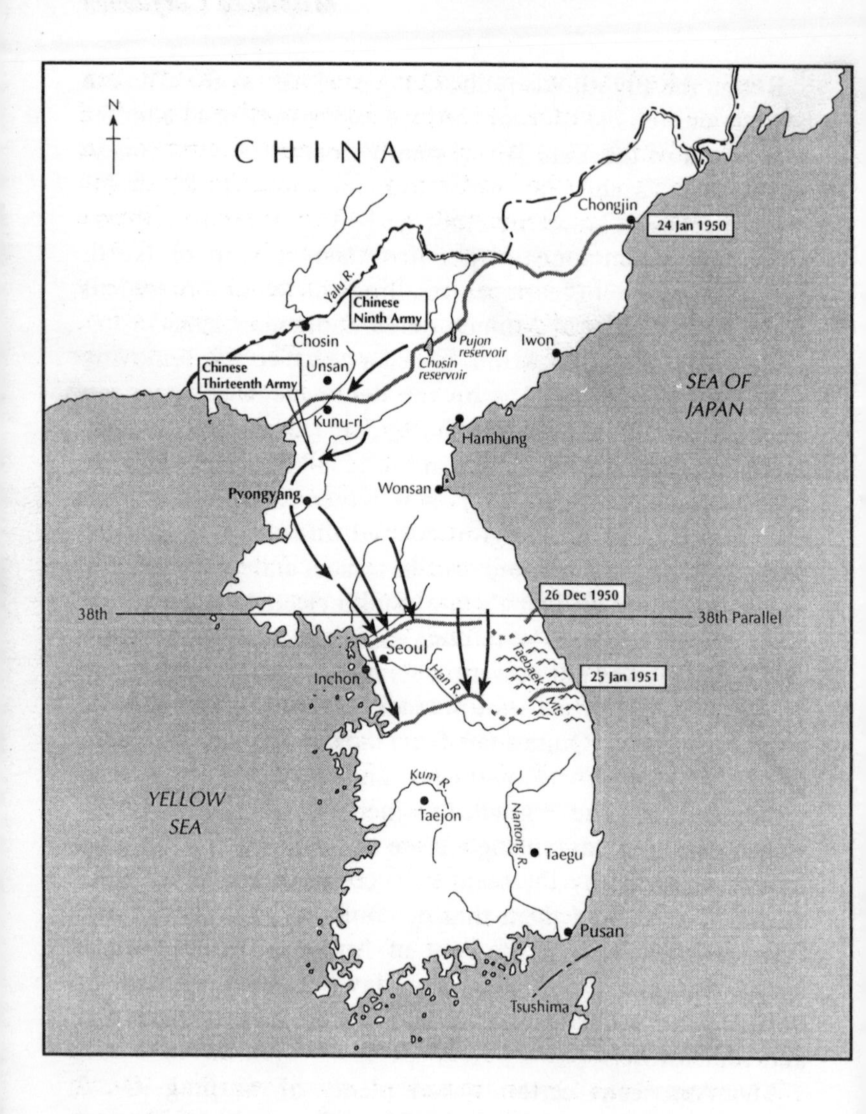

N
CHINA
Chongjin
24 Jan 1950
Yalu R.
Chinese Ninth Army
Chosin
Pujon reservoir
Iwon
Chosin reservoir
Chinese Thirteenth Army
Unsan
SEA OF JAPAN
Kunu-ri
Hamhung
Wonsan
Pyongyang
26 Dec 1950
38th
38th Parallel
Seoul
Taebaek Mts
Inchon
Han R.
25 Jan 1951
Kum R.
YELLOW SEA
Taejon
Nantong R.
Taegu
Pusan
Tsushima

The first ROK troops reached the Yalu River on 25 October. That same day, the ROK II Corps, moving north on the left of the UN advance, was attacked and almost destroyed. Its assailants were identified as Chinese. 'Are there many of you here?' General Paek, commanding II Corps, asked a Chinese prisoner of war. 'Many, many,' came the reply.

Passed on to the Americans, this intelligence was largely ignored. General Walker, commanding the US Eighth Army, sought to explain away the presence of Chinese troops by pointing out that 'a lot of Mexicans live in Texas'.

But alarm bells continued to ring. On 1 November, having passed through ROK troops near Ansung, about halfway across the peninsula, elements of the US 1st Cavalry Division were surprised by Chinese forces and forced to retire. Further east, the ROK I Corps had already been halted by the Chinese Forty-Second Army on the road to the hydro-electric plants of the Chosin Reservoir. Even the Argyll and Sutherland Highlanders from the British 27 Brigade had lost six killed and five wounded in a skirmish with Chinese troops near the Chongchon River. 'They were unlike any enemy I had seen before,' wrote Lieutenant Colin Mitchell. 'They wore thick padded clothing, which made them look like little Michelin men. I turned one body over with my foot, and saw that he wore a peaked cap with a red star badge. These soldiers were Chinese.'

Yet still the Americans could not accept that the Chinese had entered the Korean War in force, even when, on 8 November, the Central Intelligence Agency (CIA) reported that there were 30,000–40,000 Chinese troops already in Korea, and a further 700,000 poised across the border in Manchuria. Washington remained convinced that Peking would not or could not operate independently of Moscow (at this date relations between the two vast communist countries were still good).

Confirmation was seemingly provided when the Chinese

broke off their immensely successful action on 6 November. Peking later claimed that having already given the Americans due warning of their will and capability to intervene, they were prepared to wait on events. But MacArthur preferred to interpret the Chinese withdrawal as evidence that they had shot their bolt. The UN drive for the Yalu would continue.

On 25 November, having celebrated Thanksgiving the day before, the UN troops resumed their advance. But as the lead companies of the US 2nd 'Indianhead' Division, part of the Eighth Army's IX Corps, were moving through hills on the east bank of the Chongchon River, they were attacked by hordes of Chinese infantry. Before long, Chinese assault groups had smashed through their forward defences and were overrunning gun lines and rear areas. By the night of the 26th, the division had been pushed back 2 miles down the Chongchon. On its left, the 25th Division was also falling back.

'There was a complete loss of leadership,' recalled Private Scarletta of the 35th Infantry Regiment. 'It was a nightmare, really. Many times, I felt that we'd never make it out of there, that to survive this would be a miracle. Those Chinese were just fanatics – they didn't place the value on life that we did.'

Yet these reverses were minor in comparison to the disaster taking place on the right of the Eighth Army. The whole of the ROK II Corps was in headlong retreat, abandoning guns, vehicles and equipment, and leaving an 80-mile gap in the Allied line between the Eighth Army in the west and US X Corps in the east. In all, 18 divisions of the Chinese Thirteenth Army Group – more than 180,000 men – had been committed. With the Eighth Army in danger of being cut off, Walker ordered its immediate retreat.

Colonel Freeman was one American officer who was convinced that the first couple of days had been a test of American strength: 'Then they found what a thin line we had, how easily the South Koreans cracked. They saw what a pushover we were – that we would not even bomb across the Yalu. Then they became very aggressive, vey bold – and stayed that way.'

Too aggressive for the US 2nd Division, trapped near Kunu-ri on the west of the peninsula. At 1.30pm on 30 November, the divisional commander, General Keiser, ordered his men to break out to the south. But the huge nose-to-tail convoy had to run a 6-mile gauntlet of enemy fire and was cut to pieces. With the road blocked by blazing vehicles, many men sat motionless, incapable of even returning fire. When night came, Chinese infantry closed in for the kill. Only a few officers, like Colonel Skeldon of the 38th Infantry Regiment, kept their heads and led their men to safety. In just one afternoon, the 2nd Division lost 3,000 men and almost all its transport and equipment. It would be months before this famous formation was considered capable of fighting again.

Pyongyang was abandoned on 5 December. The Eighth Army had suffered 11,000 casualties in the first few days of the Chinese offensive and was still in headlong retreat. Only X Corps, to the east of the central spine of mountains in North Korea, was able to salvage some honour from one of the most inglorious episodes in American military history. Ferociously attacked on 27 November by more than 100,000 Chinese troops, its two divisions – the 1st Marine and 7th Infantry – performed a classic fighting withdrawal down the road from the Chosin reservoir towards the port of Hamhung.

None performed more heroically than the 10,000 men of the 5th and 7th Marine Regiments, who on 1 December began an epic 14-mile retreat down the west side of the reservoir from Yudam-ni to Hugaru. Their leapfrogging battalions had to clear every pass as they went. The American public, hungry for good news, was delighted, and never more so than when General Smith, commanding the 1st Marine Division, told war correspondents on 4 December: 'Gentlemen, we are not retreating. We are merely advancing in another direction.'

On 10 December, the first marines arrived at Hamhung. They had sustained 4,400 battle and 7,300 non-battle casualties (most of the latter were minor cases of frostbite). The Chinese are said to have lost 37,000 men, many from the cold. But victory was theirs. On Christmas Eve 1950, the seaborne evacuation of X Corps and its ROK allies – 100,000 men in all – from Hamhung to Pusan was completed.

Meanwhile, the Eighth Army struggled back beyond the 38th Parallel. On 4 January 1951, Seoul was captured by the Chinese. Only at the end of the month, thanks to the vigorous leadership of General Matthew B. Ridgway – who had succeeded Walker, killed in a car crash, in late December (and who had commanded US 101st Airborne Division in Operation Market-Garden) – was the line finally stabilized deep inside South Korea.

In April, MacArthur was relieved of his command and replaced by Ridgway. But the war still dragged on for another two years and ultimately claimed the lives of 450,000 South Koreans, 33,000 Americans and 3,000 servicemen from other nations, including Britain. China and North Korea lost an estimated 1.5 million men. An armistice was finally signed in July 1953; the terms of the subsequent peace treaty simply re-established the pre-war borders of North and South Korea. Such an outcome could have been achieved before Christmas 1950 if MacArthur and the US government had not been so ready to underestimate the political will and fighting capability of the Chinese.

Dien Bien Phu

On 7 May 1954, after a fifty-five day siege, the French-held mountain stronghold of Dien Bien Phu in north-east Vietnam fell to Ho Chi Minh's communist troops. It marked the end of the nine-year Indo-China War, sounded the death knell for France's Far East empire and led to the partition of what became Vietnam – which in turn resulted in the futile American attempt to prolong the existence of the corrupt non-communist regime in South Vietnam. Once again, an arrogant commander had underestimated the capabilities of his Asian foe.

Before the Second World War, French Indo-China was made up of five countries: Cochin China, Tonkin, Annam, Cambodia and Laos, as they were then called, the first three collectively making up Vietnam. When the war ended with Japan's surrender in August 1945, the French assumed that they would be able to reassert their pre-war colonial rule. But Japanese successes against European armies in the early part of the war had dispelled the notion of white superiority, and encouraged self-determination among the indigenous peoples.

In Vietnam, the most effective resistance to the Japanese during the war had been provided by Ho Chi Minh's communist-dominated Viet Minh. On 2 September 1945, Ho filled the political vacuum left by the departing Japanese by seizing Hanoi and proclaiming the independent Democratic Republic of Vietnam.

Paris, however, was not prepared to accept this *fait accompli* and French troops began to arrive in Saigon in November 1945. With the Viet Minh particularly strong in the north, they concentrated on securing the southern half of the country, together with Laos and Cambodia. Towards the end of the 1946, the French government offered Ho limited independence under the nominal leadership of the Emperor Bo Dai. Ho refused, French troops occupied Hanoi and Haiphong, and the war began in earnest.

By March 1947, the French had gained control over most of the urban areas in the north, while the Viet Minh had withdrawn to the mountainous area near the border with China. Over the next few years the French concentrated on securing the towns and lowlands of the north, with only the occasional foray into the hinterland. This enabled Ho's forces to recover and even increase their strength; they benefited greatly from the assistance of China's communist rulers after their victory over Chiang Kai-shek's Nationalists in 1949.

By the summer of 1950, with four well-trained divisions at his disposal, 40,000 troops in all, Ho's military commander, General Vo Nguyen Giap, felt strong enough to go on the offensive. A series of surprise attacks cleared the French out of northern Tonkin province and even threatened Hanoi and Haiphong. The French responded in December by sending out the experienced General Jean-Marie de Lattre de Tassigny, who immediately ordered the construction of a chain of defences in the Tonkin Delta area. Repeated attempts to breach them in the first half of 1951 resulted in Viet Minh casualties of 20,000.

Riddled with cancer, de Lattre de Tassigny returned to France at the end of the year (he died in 1952), and was replaced with the more cautious General Raoul Salan. He made little effort to regain northern Tonkin and was content to stand on the defensive. Once again, Giap took advantage of the breathing space to increase his forces to six divisions, backed by strong artillery support.

In the autumn of 1952, the Viet Minh overran the area between the Red and Black Rivers. The French counter-attack – Operation Lorraine – was a failure but this made Giap overconfident. Repeated assaults on the French garrison at Na San in December were bloodily repulsed. Giap had more success the following April when he invaded neighbouring Laos. But his troops soon ran out of supplies and were forced to withdraw, leaving behind a small force to harry the French.

Including Vietnamese troops, the French now had 175,000

men in the region; but up to 100,000 were on essentially defensive duties. The Viet Minh, by contrast, had a regular army of 125,000 men, supported by 75,000 regional troops and up to 350,000 largely untrained militiamen. The war was slipping away from the French.

In an attempt to reverse this trend, General Henri Navarre was appointed Commander-in-Chief in May 1953. A cavalryman who had spent seven years with army intelligence, his active service had been restricted to fighting nationalist rebels in Syria, Morocco and Algeria. He had, from this experience, developed a contempt for the military capabilities of 'inferior' races that was to cost him dear in Vietnam.

Within two months of arriving in Saigon, Navarre had devised a new strategy to defeat the Viet Minh. It involved dividing the region into two zones: north and south. In the north the French would remain largely on the defensive, carrying out a pacification campaign in the Tonkin Delta and launching the occasional attack to disrupt Giap's operations. In the south, where the Viet Minh were weakest, they would conduct offensives in both Annam and the Central Highlands. At the same time they would build up the Vietnamese Army and bring in reinforcements to create a mobile force large enough to engage and defeat Giap in the field.

One of the main elements of his new strategy was the reoccupation of Dien Bien Phu, a small mountain outpost on the border between north-west Vietnam and Laos. It would then become an air-supplied 'hedgehog' from which Viet Minh forces advancing into Laos could be intercepted. If Giap chose to attack it, Navarre estimated that the Viet Minh would not be able to bring to bear more than two light divisions and minimal artillery. Such a force could easily be repulsed.

Operation Castor was launched on 20 November 1953, when two parachute battalions were dropped on Dien Bien Phu, taking its Viet Minh defenders by surprise. The fierce house-to-house fighting left 110 Viet Minh and 14 French soldiers dead. More troops and equipment landed that day and the next, bringing the garrison strength up to six parachute battalions, two artillery batteries, sappers and mortars. They were soon joined by 700 Vietnamese troops who had just been evacuated from nearby Lai Chau. The French now set about constructing two airfields and a chain of nine strongpoints around the village.

When Navarre received intelligence that elements of no less than four Viet Minh divisions (one with heavy artillery) were converging on Dien Bien Phu he was unconcerned. In his opinion, they were heading towards Laos and he rejected the proposal by General Cogny, the area commander, to launch strikes into the communist heartland of the Viet Bac to forestall a move against Dien Bien Phu.

Even if the French outpost was their target, Navarre commented in a directive of 3 December, Giap's divisions would need six weeks' 'movement', a recce phase of six to ten days, and a battle lasting 'several' days, ending in a Viet Minh defeat. The reality was to prove very different.

'Navarre underestimated the fighting and logistical capabilities of Asian troops,' wrote John Colvin, Giap's biographer, 'just as Wavell had underestimated the Japanese in 1942. Both their armies suffered in pain and humiliation for their generals' ignorance, but unlike Wavell, Navarre did have intelligence telling him that his resources were wholly insufficient to resist the enemy forces about to overwhelm his regiments.'

Before long, the converging Viet Minh forces had closed the ring round Dien Bien Phu and heavy artillery was ranging on its key points. When French parachute commando units went on reconnaissance patrols, they were invariably ambushed. Though casualties were in the region of 1,000 men by the end of February, virtually nothing was done to improve the defences against the inevitable communist attack. Even ammunition supplies were not replenished for a whole week 'because of other priorities'.

By 13 March 1954, the day Giap launched his first major assault, Dien Bien Phu was overlooked from its surrounding hills by a force of four infantry divisions – 50,000 men in all, with 31,000 support troops and 23,000 more further back. In addition, the Viet Minh had more than 200 artillery pieces (almost half were 105mm calibre or greater) in well-camouflaged positions. The French garrison numbered 13,000 men, of whom less than 7,000 were combatant, with only 60 guns of greater calibre than 57mm.

Navarre had gravely underestimated his 'enemy's ability to bring in guns and ammunition along primitive jungle tracks, through the rains, in Russian trucks, on thousands of bicycles, in innumerable sampans and bamboo rafts, with convoys of pack horses and hundreds of thousands of porters', wrote Colvin. Such was the physical hardship of this heroic effort that Giap later estimated that, in order to transport 1 kilo of supplies, each porter required 21 kilos of supplies to sustain him.

On the first day of the attack, Viet Minh artillery bombarded the main airfield, destroying fighter-bomber and reconnaissance aircraft on the ground. The last supply plane arrived that evening and the subsequent air drops were hopelessly inefficient thanks to the severity of Giap's flak. During the night of the 13th, strongpoint Béatrice on the north-east perimeter, manned by part of the illustrious 13th Foreign Legion, was stormed and taken. Three hundred Legionnaires were killed; all the officers were either severely wounded or dead. Colonel de Castries, the garrison commander, chose not to counter-attack because he wanted to preserve troops for strongpoint Gabrielle in the extreme north.

Gabrielle fell the following day, as did strongpoint Anne-Marie, just above the northern airstrip, a day later. Some 2,500 Viet Minh are said to have lost their lives at Gabrielle, as against 90 of the 500 Algerian defenders. The prisoners were marched out over the bodies of communist soldiers strewn over the barbed wire. One was still living, his guts hanging out. 'Step on him,' said a Viet Minh officer. 'He has done his duty for the People's Army.'

In three days, the enemy had taken a third of the garrison's nine strongpoints and it was now impossible to evacuate the wounded. In December, Colonel Piroth, the garrison's artillery commander, had told Navarre: 'Mon général, no Viet Minh cannon will be able to fire three rounds before being destroyed by my artillery.' Now he preferred death to dishonour and blew himself up with his own grenade. Also that day, de Castries's Chief of Staff had a nervous breakdown.

There was now a lull as Giap brought in reinforcements to replenish his heavy losses. The French, too, parachuted in more troops and supplies, although much of the latter fell behind enemy lines. On 30 March, the attack resumed against Dominique and Elaine, the two strongpoints to the east of the village. Half of each were overrun, as was part of Huguette on the west side. The following day, Colonel Langlais, commanding the airborne forces, took over effective command from de Castries, who was seeking an honourable surrender. 'It was as if a spring had broken inside him.'

This change stiffened the defenders and on 10 April more reinforcements were parachuted in; unfortunately, 40 per cent of the 850 men dropped straight into Viet Minh hands. But enemy casualties were severe – 10,000 since the start of the siege – and on 11 April Giap gave up frontal attacks in favour of a battle of attrition. Soon after, the monsoon broke, hindering air resupply and giving cover to the Viet Minh as they wore down the remaining defences. General Cogny now suggested sending in a relief force, but Navarre was unenthusiastic.

By the end of April, the French were out on their feet. On half rations of rice and fish sauce, they were often up to their knees in water in the flooded trenches. Their defences were beginning to collapse under the twin impact of monsoon and artillery fire. The noise of the guns and the insanitary living conditions were almost unbearable. After 50 days of fighting, French effectives were down to 2,900; the Viet Minh still had 30,000.

On 1 May, Giap launched the final phase of the battle. A huge artillery bombardment was followed by a massive two-division infantry assault against the surviving positions in Dominique and Elaine. Despite gallant Foreign Legion counter-attacks, more and more posts were lost. Badly wounded men returned to the front line from the field hospital. 'If we've got to go, we'd rather go with our mates,' many were heard to say.

Still in nominal command, de Castries reported to Hanoi: 'No more resources left, extreme fatigue and weariness of all concerned.' Cogny replied that there should be 'no capitulation nor rout' but left it up to de Castries to decide on a break-out. Despite desperate resistance, only isolated pockets of resistance remained by 7 May. At 10am, de Castries told Cogny that they were still 'holding on, tooth, claw and nail'. Cogny replied that the battle 'had to be finished now', but without capitulation or white flag. The last message received from Dien Bien Phu was: 'We're blowing up everything. Adieu.' In full-dress uniform and medals, de Castries surrendered to General Vu of the 308th 'Iron' Division at 5pm.

The Feast of Camerone

At 10pm on 30 April 1954, just one week before he surrendered the besieged French garrison of Dien Bien Phu to the Viet Minh, Colonel de Castries reported a

cont. overleaf

minor victory to his superiors in Hanoi. Major Coutant's 1/13th Foreign Legion had carried out a successful raid on Viet Minh trenches and fortifications south of the Eliane 2 strongpoint. One bunker had been completely destroyed, two others severely damaged and at least ten enemy soldiers killed.

What the communiqué failed to mention was that the raid had been launched with the intention of retrieving two crates of 'Vinogel' wine concentrate from no man's land so that the Legionnaires could celebrate the anniversary of their most glorious action: the Battle of Camerone. In that epic encounter in 1863, a single company of Legionnaires, just 65 men in all, had held off 2,000 Mexicans for more than 12 hours, inflicting 800 casualties in the process.

As there was only one bottle of wine per platoon in Eliane 2, the Legionnaires were not about to let the 'Vinogel', manna from heaven, fall into Viet Minh hands. So a raiding party of volunteers was organized (*everybody* would have volunteered for that task, observed one non-Legionnaire) to collect the wine after dark. Knocking out the enemy bunkers was merely a subsidiary to the main objective.

Another Camerone tradition was for the senior officer to create honorary Legionnaires. Each person so honoured is sponsored by a Legionnaire who acts as his 'godfather' and whose serial number, followed by the suffix *bis* (twice), would become his own. That morning, one of those honoured was Geneviève de Galard, an Air Force nurse who had been stranded when her ambulance plane was blown up on the runway. Her unflinching work under appalling conditions in the main field hospital had already been recognized with the award of the Croix de Guerre and the Légion d'Honneur.

Now the heroic young woman, whom the rear-echelon press corps had nicknamed the 'Angel of Dien Bien Phu', was made an honorary private in the Foreign Legion. After the ceremony, she turned to her 'god-father', an officer's batman, and said: 'If we ever get out of this alive, I'll pay you a bottle of champagne no matter where we meet.'

In 1963, driving in Paris with her husband, Geneviève recognized the Legionnaire walking along the pave-ment. She stopped immediately, embraced him and made good her promise.

Since the start of the battle, the French had suffered about 9,000 casualties, 2,000 of them fatal. About 8,000 men and women (including some North African prostitutes) were taken prisoner, but only 3,000 saw their homes again. Viet Minh casualties were in the region of 23,000. When Giap gave the news to Ho Chi Minh, the latter replied: 'However great the victory, it is only the beginning.' Prophetic words in view of the fact that it would take until 1975 before Vietnam was finally united under communist rule.

But for the French the battle was fatal. Within months, a ceasefire had been negotiated at an international conference in Geneva. The French and their Vietnamese allies agreed to move south of the 17th Parallel, leaving the communists in control of what came to be known as North Vietnam. Two years later, elections were held in the south and France officially withdrew from its Far East empire. It would have happened anyway, but the process was undoubtedly accelerated by Navarre's serious underestimation of his enemy's capabilities.

Chapter 5

A Failure to Perform

Despite appearances, not all military blunders are the fault of senior commanders and politicians. Some of the more spectacular setbacks can be put down to a failure of duty on the part of the ordinary soldier and his officers. This can manifest itself as indiscipline, lack of initiative or, as is most often the case, poor fighting resolve.

Those who fall into the latter category tend to cling to the maxim: 'He that fights and runs away, may live to fight another day.' Certainly, one indication of a spineless performance is when the proportion of prisoners of war to overall casualties is unusually high. But capitulation is not always the safest course of action – as the Spanish learnt to their cost in Morocco.

Low morale and inexperience are two reasons why troops perform poorly. The quality of the opposition is another. It may be no coincidence that in four of the six examples given here, the victorious troops were German. On the other hand, and contrary to popular belief, Italian soldiers are not the only ones to underperform on the battlefield. As will be shown, French, Spanish, British, New Zealand and American troops have all belied their vaunted military reputations at one time or another.

Crécy

The French defeat at Crécy, on 26 August 1346, by King Edward III's numerically inferior English army was due in no small part to the effectiveness of his longbowmen. Even more decisive, however, was the indiscipline of the French knights and the gutless performance of their crossbowmen, who were Genoese mercenaries.

Edward was just fourteen when he became King of England on 15 January 1327, two days after the deposition of his dissolute father, Edward II. The following year, his maternal uncle, King Charles IV of France, died and was succeeded by his Valois cousin, Philip VI, on the grounds that the crown could only be transmitted through the male line. Edward's mother, Isabella, immediately objected because her son was a descendant of the senior branch of the Capetian line (the Capetian kings ruled France from 987–1328, Charles IV being the last. The Valois dynasty succeeded to the throne). At first the claim was just talk, for the young king was in no position to take any action, but it later became useful as a legal justification for what became the Hundred Years' War.

The real reason for the conflict, however, was that Edward was also Duke of Aquitaine – thanks to the marriage of Eleanor of Aquitaine to his ancestor, Henry II – and therefore the theoretical vassal of the French monarch; the Duchy of Aquitaine was by the same token an English possession. But as a king in his own right, Edward refused to behave like an inferior. In 1336, after Philip had threatened to intervene on the side of the temporarily deposed King David II of Scotland, Edward mobilized his troops and ordered his fleet to concentrate in the English Channel. He had already annoyed his liege-lord by receiving Robert of Artois, a man Philip VI had banished for poisoning his wife, the king's sister. Philip now responded to both these provocations by sending an army to the borders of the English-held province of Guienne – the territory remaining from the old Duchy of Aquitaine (the other province, Gascony, being by now in French hands) – and declaring it forfeit.

It was only now, according to the French historian, Professor Perroy, that Edward 'assumed the position of claimant to the throne of the Capetians' in order to transform 'the feudal conflict, in which he was an inferior, into a dynastic struggle, which would make him his adversary's equal'. On 1 November 1337, therefore, he sent the Bishop of Lincoln to Paris with an ultimatum addressed to the *soi-disant* (in other words, usurper) King of France.

To conquer the kingdom of France was out of the question. Though smaller than the France of today, it was much bigger and more prosperous than England, with a population five times as great. In addition, it was surrounded by a number of semi-independent duchies and counties that paid homage to its ruler and were supposed to provide troops in time of war. Edward therefore confined himself to distracting Philip from Guienne by attacking France from the north.

In need of allies, he used money and family connections to win over, among others, the Emperor Ludwig of Bavaria, the Duke of Brittany, the Count Palatine of the Rhine, and the rulers of Hainault, Limburg and Brabant. But the powerful Count of Flanders stayed loyal to Philip and, to prove it, put a strong garrison on the island of Cadsand, off Sluys, to hamper English communications with the Low Countries. When an English expedition landed on the island and captured the post on 11 November 1337, the Hundred Years' War officially began (although it would be 114 years before it ended).

The subsequent skirmishes were inconclusive. But on 24 June 1340, Edward gained the upper hand when his ships destroyed the French fleet outside the Flemish harbour of Sluys. Tactically naive, the French had remained in a dense mass, their foremost ships linked together with iron chains. This enabled the English to reduce them one by one: longbowmen would keep the heavily defended top castles busy with showers of arrows, while men-at-arms boarded the ships and took control in hand-to-hand fighting. The French lost 166 vessels and many lives. So much Gallic blood was drunk by the fish, it was said afterwards, that had God given them the power of speech they would have spoken in French.

No one dared give Philip VI the bad news. Pushed forward eventually, his jester said, 'Oh, the cowardly English, the cowardly English!' Asked why, he replied, 'They did not jump overboard like our brave Frenchmen.' The king got the point.

Though Edward failed to take advantage of this crushing victory – his senseless siege of Tournai had to be abandoned after two months because there was not enough money to pay his troops – it guaranteed him control of the Channel for more than a generation. This in turn gave his quarrel with Philip new impetus.

In July 1346, two years after the Houses of Parliament had voted him subsidies to end the war 'by dint of the Sword', Edward was ready to undertake another campaign against Philip. It was an auspicious moment because, in response to the capture of Bergerac by an English army the previous autumn, the French, led by Philip's son, the Duke of Normandy, were busy trying to reduce the powerful castle of Aiguillon in Guienne. It also made sense, with the Duke and his men absent, to strike for Normandy.

Guided by Geoffrey de Harcourt, a Norman baron banished from France, and accompanied by his own eldest son, the seventeen-year-old Edward, Prince of Wales, King Edward landed at La Hogue in the Cotentin peninsula on 12 July. It took six days to disembark his army of 3,000 men-at-arms (or knights), 5,000 archers and 1,500 pikemen – a total, with supernumeraries, of around 10,000 men.

Divided into three columns, or 'battles', the army left La Hogue on 18 July, burning and pillaging as it went. It met its first serious opposition at Caen eight days later. Philip had sent his Constable, the Count of Eu, with a force of knights to organize Caen's defence. But the citizens, who had insisted on fighting in the open, took flight at the first sight of the English and the unwalled town was ruthlessly sacked. The Constable and many knights were captured and sent back to England.

Edward also ordered the wounded and a huge haul of booty to be embarked in the fleet, then at the mouth of the River Orne, but before this could take place the crews mutinied and set sail for home. This placed the army in an invidious position. Without a line of communication it could not stay where it was. To move south was out of the question because the Duke of Normandy, with his larger army, was already on his way north. The only option was to move east towards Flanders where the English could be sure of ships.

Finding the bridge at Rouen destroyed, Edward moved up the Seine, searching for a crossing. All the while, Philip and his ever-growing army shadowed them from the opposite bank. At Poissy, just 20 miles from Paris, the bridge was only partially destroyed and Edward paused for five days while it was rebuilt. Philip made no attempt to dispute the crossing, which took place on 16 August, and instead fell back to St Denis to await reinforcements.

At Airaines, 10 miles south of the River Somme, Edward received word that all the bridges and fords from Abbeville upstream were either broken or held in force. By now, Philip was at Amiens, only a short way upriver, and his army was growing stronger by the day. There was not a moment to lose. Deliverance came in the form of a prisoner from Mons-en-Vimeu, Gobin Agache, who offered to guide the English to a ford over the Somme at Blanque Taque ('the White Spot'), 10 miles below Abbeville, in return for a reward and his freedom.

Marching at night, Edward reached the ford early on 24 August. An hour or so later, the tide went out and his army waded across, easily defeating a strong French force that had gathered on the opposite bank. Philip was now rapidly approaching from the south, only to hear that the English had crossed and the tide had come in. He hurried east to use the bridge at Abbeville.

That night, Edward slept at Noyelles, and continued his march the following morning toward the forest and village of Crécy-en-Ponthieu in Picardy. There, according to the contemporary French chronicler Jean Froissart, he told his men: 'I will take up my position here and go no further until I have a sight of the enemy. I have good reason to wait for him, for I am on the land I have lawfully inherited from my royal mother, which was given to her as her marriage portion. I am ready to defend my claim to it against my adversary Philip of Valois.'

A more practical reason was that, with the French army less than a day's march away, a battle was likely and it made sense to choose the most advantageous site possible. Assuming that Philip would advance up the Abbeville–Hesdin road, and knowing that chivalrous etiquette precluded anything other than a frontal attack, Edward selected a position that best suited the tactics of his army. It was on a broad hill between the villages of Crécy and Wadicourt, with the ground in front gradually falling away to the village of Estrées-les-Crécy, a little to the west of the Abbeville–Hesdin road.

Edward placed two of his three battles on the forward slope of the rise, and the other slightly to the rear on the Crécy–Wadicourt road. The right forward battle, made up of 800 dismounted men-at-arms and 1,000 spearmen, was nominally commanded by the Prince of Wales; the actual commanders were Warwick, the Earl Marshal, and the Earls of Oxford and Harcourt. The left battle, made up of just 800 men-at-arms, was commanded by the Earl of Northampton, the Constable, and the Earl of Arundel. Edward himself commanded the rear battle of 700 mounted men-at-arms, with his headquarters at a windmill on the Crécy rise.

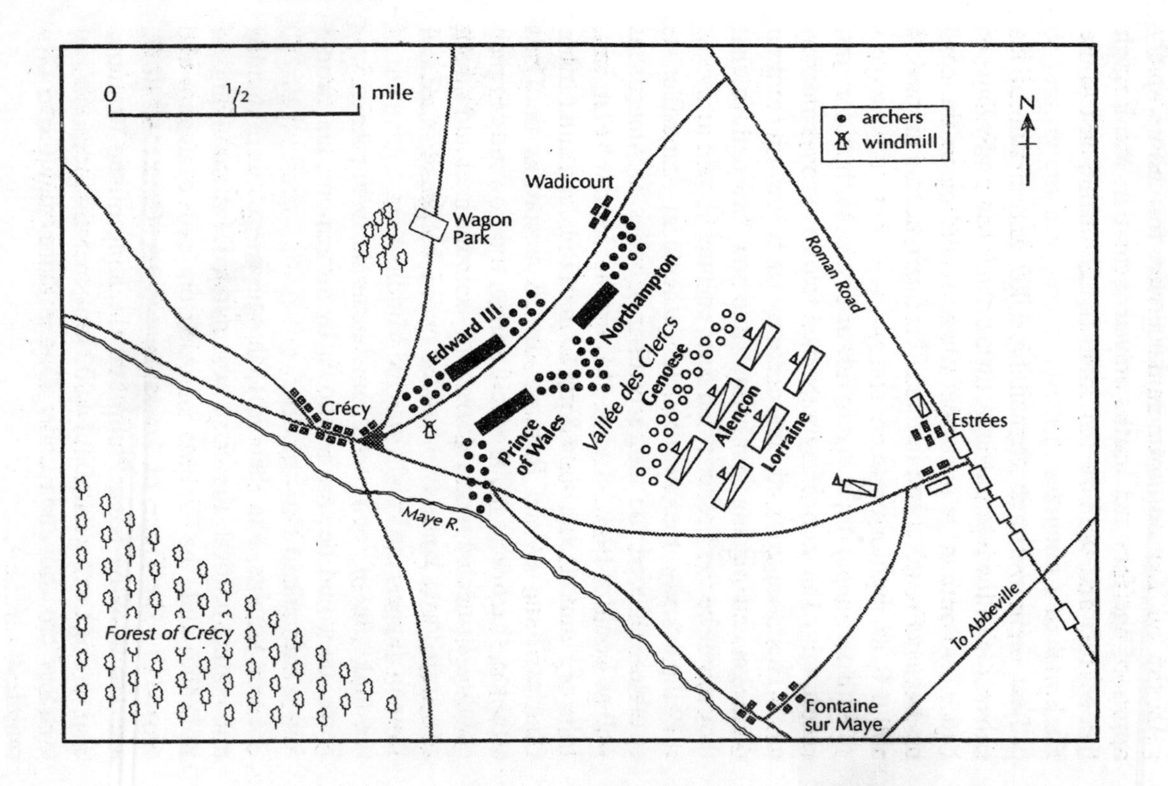
0
1/2
1 mile
archers
windmill
N
Wadicourt
Wagon Park
Roman Road
Northampton
Edward III
Vallée des Clercs
Genoese
Crécy
Prince of Wales
Alençon
Lorraine
Estrées
Maye R.
Forest of Crécy
To Abbeville
Fontaine sur Maye

Of the 5,000 or so bowmen, 3,000 were formed up on the flanks of the forward battles *en herse*; that is, at an angle thrown forward, with the two inner flanks meeting in a point. They had dug ditches and driven stakes into the ground to protect their front. The remaining 2,000 were drawn up on either side of the rear battle. A little further back still, Edward had 'enclosed a large park near a wood', and placed in it 'all his baggage wagons and horses'. The English army numbered about 8,500 fighting men.

With the deployment over, Edward rode along the ranks, encouraging the men. They were then told to eat their midday meal. Their hunger sated, they re-formed in battle order and sat on the ground 'placing their helmets and bows before them, that they might be the fresher when their enemies should arrive.'

Philip's army, meanwhile, had gathered at Abbeville. Its notables included the king's brother, Charles of Alençon as well as Louis of Blois, Rudolf of Lorraine, Louis of Flanders, John of Hainault, the blind King John of Bohemia and his son Charles, King of the Romans. In all, according to Lot's estimate, there were 8,000 men-at-arms, the flower of French nobility. Protected by plate armour, visored helmet and shield, and armed with lance, two-handed sword, battleaxe, mace and 18-inch dagger, the knight was the Middle Ages' equivalent of the tank. But if separated from his similarly burdened war horse, he would be weighed down by his armour, and would find it difficult to rise unaided.

These knights, like their English equivalents, were divided into three battles: the first commanded by the King of Bohemia and the Counts of Alençon and Flanders; the second by the Duke of Lorraine and the Count of Blois; and the third by King Philip and the King of the Romans. Supporting them were about 4,000 foot, including a number of Genoese crossbowmen under Odone Doria and Carlo Grimaldi.

Unsure of where the English were, Philip sent four knights ahead to reconnoitre. One of them, Le Moine de Bazeilles, reported that Edward was in a strong position near Crécy, and suggested that the King should halt his vanguard to enable the rest of the army to catch up. They could then attack as a single body the following morning. Philip agreed and the order was given to the standard-bearers: 'Halt banners on the King's orders, in the name of God and St Denis.'

'At this command,' writes Froissart, 'the leaders halted, but those behind continued to advance, saying that they would not stop until they had caught up with the front ranks. And when the leaders saw the others coming they went on also. So pride and vanity took charge of events. Each wanted to outshine his companions, regardless of the advice of the gallant Le Moine.'

With neither Philip nor his marshals able to restrain them, 'for there were too many lords among them, all determined to show their power', they rode on 'in no order or formation'. But when their leaders 'saw the English they reined back like one man, in such disorder that those behind were taken by surprise and imagined they had already been engaged and were retreating'. The confusion was added to by the many volunteers from the district who 'crowded the roads between Abbeville and Crécy, and when they came within ten miles of the enemy they drew their swords and shouted: "Kill! Kill!" Yet they hadn't seen a soul.'

Arriving in the midst of this disorder, Philip realized that a battle could not be avoided and so ordered his Genoese crossbowmen to advance. But having just marched 18 miles, burdened by armour and the weight of their cross-bows, they 'would sooner have gone to the devil than fight at that moment', and told their commanders so. This caused the Count of Alençon to rage: 'What is the use of burdening ourselves with this rabble who give up just when they are needed!'

It was after 4pm when the Genoese were finally persuaded to advance. But no sooner had they loosed the first shots than 'the English archers took one pace forward and poured out their arrows on the Genoese so thickly and evenly that they fell like snow,' recorded Froissart. 'When they felt those arrows piercing their arms, their heads, their faces, the Genoese, who had never met such archers before, were thrown into confusion. Many cut their bowstrings and some threw down their crossbows. They began to fall back.'

Seeing this, Philip called out to the men-at-arms of his first battle: 'Quick now, kill all that rabble. They are only getting in our way!'

Thereupon, these splendidly mounted knights began to struggle through the fleeing Genoese, stabbing with their lances and hacking with their swords as they went. But the English continued to shoot their arrows into the thickest part of the crowd, impaling and wounding horses and riders, 'who fell to the ground in great distress, unable to get up again without the help of several men.'

The Longbow

The ace in Edward III's pack during the Battle of Crécy was undoubtedly the longbow. A Welsh invention, it had been developed by his grandfather, Edward I, for use against the Scots in the Highlands. Made of yew, 6 feet long, it could fire up to 12 arrows a minute in skilled hands. Deadly accurate at 200 yards, its 3-foot iron-tipped arrows were capable of carrying another 100 yards if necessary. The effect of its lethal hail was at the same time destructive and demoralizing.

So crucial was the longbow to Edward's chances of success in the war with France that in 1337 he outlawed, on pain of death, all sport except archery, and cancelled the debts of all men who made the bows and their arrows.

The French also used archers and crossbowmen, generally Genoese mercenaries. But they refused to give them the tactical scope allowed by the English because this would diminish the status of their knights. Chivalry maintained that horsemen should engage their equals in close combat. According to a twelve-century song, the first archer was 'a coward who dared not come close to his foe'.

Nevertheless, the crossbow was still a formidable weapon. Made of wood, steel and sinew, it shot a bolt with more penetrating power than a longbow arrow. However, it was cumbersome to wield and slow to load (firing an average of two bolts a minute), and was more effective in siege conditions than in open battle. Banned by the Church in 1139, shortly after its invention, on account of its terrifying power, it had since failed to supplant the mounted knight as the premier fighting force – for a determined charge of knights would invariably shatter a line of crossbowmen.

By now, the second battle had come forward to add to the confusion. Somehow, out of this disorderly mass, the French knights launched the first of many assaults. They tended to ignore the English archers, partly because of their defences and partly because the rules of chivalry demanded that a knight could only fight his equal. Instead, they engaged the dismounted knights of the two forward English battles, particularly the Prince of Wales's. At one point the situation was so desperate there that Warwick sent Sir Thomas of Norwich to ask Edward for assistance. 'Is my son dead or stunned, or so seriously wounded that he cannot go on fighting?' queried the King.

'No, thank God,' replied Sir Thomas, 'but he is very hard pressed and needs your help badly.'

'Go back to him and those who have sent you,' said the King, 'and tell them not to send for me again today, as long as my son is alive. Give them my command to let the boy win his spurs, for if God has so ordained it, I wish the day to be his.'

As a precaution, Edward sent forward the Bishop of Durham and thirty knights. He might have been more worried for his son's safety if he had not already spotted the left battle, under Northampton, wheeling to its right to take the attacking French in the flank. As they were driven back, Philip appeared with the third French battle and the attack was continued.

In all, there were fifteen separate assaults, the last ones taking place after dark. But by then the French cause was hopeless, and many knights were killed as they wandered about the field looking for their leaders. Wounded in the neck by an arrow, Philip had already fled, urged by the Count of Hainault, who grabbed his bridle and said: 'Sire, lose not yourself wilfully.'

Accompanied by just Hainault and four other lords, he sought temporary refuge in the castle of La Broye. 'Who comes knocking at this hour?' shouted the captain of the gate.

'Open your gate quickly,' Philip replied. 'It is the unfortunate King of France.' Unlike Edward II after Bannockburn, he was not refused entry.

Not once during the battle did Edward allow his troops to pursue the retreating French. According to Froissart, only 'pillagers and irregulars, Welsh and Cornishmen armed with long knives' were allowed out after the French, 'and, when they found any in difficulty, whether they were counts, barons, knights or squires, they killed them without mercy. Because of this, many were slaughtered that evening, regardless of their rank. It was a great misfortune and the King of England was afterwards very angry that none had been taken for ransom.'

More than 1,500 French and allied knights were killed, among them the Counts of Alençon, Flanders, Blois and Sancerre, the Duke of Lorraine, the King of Majorca, and King John the Blind of Bohemia, who had been led into battle with his horse fastened to those of his knights. His crest of three ostrich feathers with the motto *Ich dien* ('I Serve') was taken by the Prince of Wales and thereafter attached to the title. The blind king's son, Charles of Bohemia, the future Emperor of Germany, was less rash and escaped. Thousands of common soldiers were killed. The English fatalities were said to be less than 100.

From Crécy, Edward marched north to the powerful fortress of Calais and set about besieging it on 4 September. It held out long enough to enable the Duke of Normandy's army to appear in July 1347. In the event, however, the French king's son thought the English position too strong and decided not to attack. Calais fell six days later, and was to remain in English hands for more than 200 years.

The defeat at Crécy, therefore, had dire consequences for the French. Yet it need never have happened. English weaponry and tactics were undoubtedly superior, but the French were numerically stronger, fighting on their own soil, and might well have prevailed if they had fought as a co-ordinated body. But thanks to the indiscipline of the arrogant French knights, the battle was fought a day too soon, the demoralized crossbowmen made little impact, and the mounted assaults were piecemeal and largely ineffective.

Caporetto

On 24 October 1917, Austro-Hungarian and German troops tore a huge hole in the Italian front line at Caporetto in Istria. Within two weeks, the demoralized Italians had surrendered more than fifty miles of ground and thousands of prisoners-of-war. To say that they failed in their duty as fighting soldiers would be something of an understatement.

Although a member of the Triple Alliance with Germany and Austria–Hungary when the First World War began in late July 1914, Italy remained neutral on the grounds that she had not been consulted about the Austrian attack on Serbia. In truth, she was anxious to see which way the wind would blow.

With both sides avidly courting her involvement, it was a case of her Prime Minister, Antonio Salandra, striking the best deal. Italy would remain neutral and not pose a threat to her southern frontier, he told the Austrians, if they ceded the southern Tyrol and territory at the head of the Adriatic, and gave autonomy to Trieste. The Austrian reply fell short in that it offered only that part of the Tyrol which contained a sizeable number of ethnic Italians. Salandra therefore began to negotiate with the Allies.

In April 1915, in return for sizeable territorial concessions, Salandra signed a secret treaty whereby Italy would enter the war on the Allied side. Once the war was over, as well as Istria and the whole of the Tyrol, Italy would receive Dalmatia, part of Albania and the Dodecanese Islands. If Turkey was defeated and partitioned, Italy was promised a share of the southern part of Asia Minor and the whole of Libya. Finally, should the German colonies there be taken over by Britain and France, Italy would be compensated in other parts of Africa. Her formal declaration of war against Austria took place on 23 May 1915 (though she would not declare war against Germany until August 1916).

Italy's border with Austria, the new front line, stretched for 350 miles from the Swiss frontier to the River Isonzo at the head of the Adriatic. Only in the Isonzo valley was the terrain less than mountainous and therefore suitable for offensive operations. Between June 1915 and September 1917, the

Italians made a series of assaults here; but these eleven battles gained little ground and cost them over a million casualties.

The low morale of the troops was not helped by the lack of support for the war back home. During the summer of 1917, the slogan of the powerful Socialist Party was 'Next winter not another man in the trenches'. Yet General Luigi Cadorna, the Italian Chief of Staff, was determined to keep up the pressure; in September 1917, he managed to persuade the British and French to lend him 200 heavy artillery pieces to support his next attack. It had been cancelled, and the bulk of the guns returned to the Western Front, when Cadorna received intelligence reports that the Central Powers were planning an offensive of their own.

With the British making minimal progress at Passchendaele (Third Battle of Ypres) and the Russians in headlong retreat, Germany at last felt strong enough to satisfy Austria's incessant demands for reinforcements with which to launch an attack against Italy. General von Dellmensingen, an expert in mountain warfare, was sent to execute a plan put forward by the Austrians in 1916. It involved attacking across the Isonzo from a small Austrian bridgehead in the Tolmino area. The intention was to drive the Italian Second and Third Armies back behind the River Tagliamento, some 30 miles to the rear.

The main assault was to be carried out by the German Fourteenth Army under General Otto von Below, consisting of seven German and three Austrian divisions. Supporting it in the north was part of the Austrian Tenth Army, and in the south the Austrian Second and First Armies. A key element of the attack was a short but fierce bombardment using gas; the Fourteenth Army alone had more than 1,800 guns and heavy mortars. Originally set for 20 October, heavy rain from the 10th forced a postponement of four days.

Initially, thanks to clever Austrian ruses, Cadorna believed that the attack would take place further west in the Trentino region north of Lake Garda. But by early October, his intelligence had identified 35 of the 53 German and Austrian divisions on the Isonzo front. Facing them were the 34 divisions of his Second and Third Armies. Clearly, the attack would be here.

This posed a severe problem for Cadorna. General Capello, the commander of his Second Army, due to take the brunt of the offensive, was all for advancing up the Vrh Valley and from the Bainsizza plateau south of Caporetto so that he could take the enemy in the flank. Consequently, Capello had concentrated all his reserves here, leaving nothing to support his left wing. When Cadorna refused to give permission for this attack, Capello neglected to alter his troop dispositions. He was sick in bed at the time, but would not agree to be relieved. When Cadorna realized that the line around and to the north of Caporetto was held by just two battalions to the mile (instead of eight elsewhere), he ordered a division to move up from the south. But it was still in transit when the attack began.

At 2am on 24 October, the Austro-German artillery opened up on a 60-mile front north of the Adriatic. Around Tolmino and Caporetto, the focal point of the assault, the two-and-a-half-hour bombardment included gas shells. The effect of this was soon felt as the lively Italian counter-fire died away. After a pause, a fresh bombardment opened at 6.30am against headquarters, ammunition dumps, approach roads and artillery positions. It was a disruptive technique that had been used with some success during the recent German victory over the Russians at Riga.

At 7am, aided by rain and mist, the attack began. The most spectacular breakthrough was made, not surprisingly, in the thinly guarded Caporetto sector. Having made a breach in the Italian line and taken Caporetto, the German 12th Division

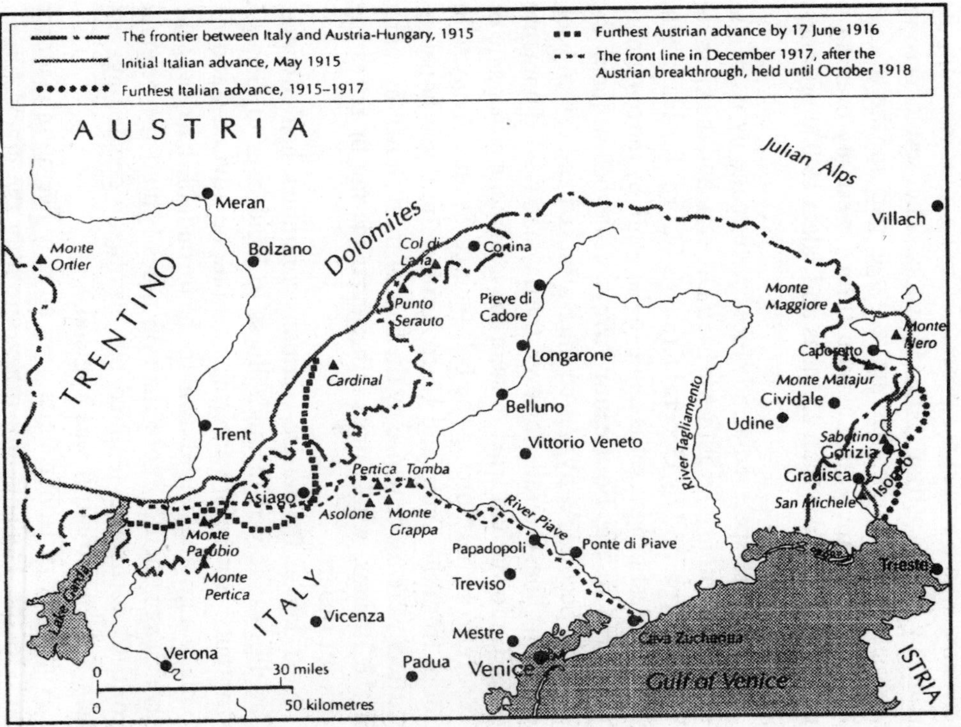

The frontier between Italy and Austria-Hungary, 1915
Initial Italian advance, May 1915
Furthest Italian advance, 1915–1917
Furthest Austrian advance by 17 June 1916
The front line in December 1917, after the Austrian breakthrough, held until October 1918
AUSTRIA
Julian Alps
Villach
Meran
Bolzano
Monte Ortler
TRENTINO
Dolomites
Col di Lana
Cortina
Punto Serauto
Pieve di Cadore
Monte Maggiore
Monte Nero
Caporetto
Longarone
Cardinal
Monte Matajur
Cividale
Belluno
Udine
Trent
River Tagliamento
Sabotino
Gorizia
Vittorio Veneto
Pertica
Tomba
Gradisca
Isonzo
Asiago
San Michele
Asolone
Monte Grappa
River Piave
Monte Pasubio
Papadopoli
Ponte di Piave
Monte Pertica
Trieste
Treviso
ITALY
Lake Garda
Vicenza
Mestre
Cava Zuccherina
Verona
Padua
Venice
Gulf of Venice
ISTRIA
0
30 miles
0
50 kilometres

wheeled right and moved north some five miles behind the Italian 46th and 43rd Divisions, which were still holding out. Then it wheeled right again, into the rear of the Italian 19th Division, which was in the process of being attacked from the front by the Austrian 50th Division. The result was its complete destruction, leaving a huge gap in the Italian front. Finally, the 12th Division turned west. By evening it had advanced more than 15 miles and taken 15,000 prisoners and 100 guns.

Slightly less spectacular advances were made by its neighbouring divisions, the 50th Austrian Division and the German Alpine Corps, and by General Krauss's I Corps further north. More modest progress was made to the south by the remaining two corps of the Fourteenth Army, while below them the Austrian Second Army was counter-attacked and driven back to its start line. Overall, however, the day had yielded immense gains. Cadorna had tried to stem the avalanche either side of Caporetto by reinforcing the beleaguered Second Army with troops from the Tenth and Third Armies. But it was hopeless. Matters were made worse by the lack of fight shown by many of the Italians, whole units either surrendering or retreating without orders.

Next day, as the Austro-Germans continued their advance, Capello was forced through illness to hand over to General Montuari. But before entering hospital, he advised Cadorna to make a gradual withdrawal to the formidable barrier of the River Tagliamento, more than 30 miles behind the original front. Cadorna was tempted but eventually decided against such a deep retreat; it was a mistake.

A number of Italian units were already fleeing in disorder back to the Tagliamento; others, notably the crack Alpini, obeyed orders and held their ground. But it was only a matter of time before they were outflanked and forced to surrender. In one such action, Captain Erwin Rommel, a young company commander in the Württemberg Mountain Battalion, captured 12 officers and 500 men.

By the end of the second day, 25 October, the Second Army was in total disarray – much of it self-inflicted. 'Reserves moving up prepared to do their duty were greeted with yells of "Blacklegs",' wrote Cyril Falls, author of *Caporetto 1917*. 'Krauss's troops encountered Italian units in formed bodies marching into captivity, calling out: "*Eviva la Austria*!" and "*A Roma*!" This picture of complete demoralization certainly did not apply to all troops, but it had made deep inroads.'

Matters soon went from bad to worse. On the 26th, Monte Maggiore, one of the intermediate positions suggested by Capello, was taken by the German Alpine Corps, as were the peaks of Kuk and Matajur to the west of Tolmino. Rommel was instrumental in both the latter actions, at one point walking towards a huge group of Italians, waving a handkerchief and calling upon them to surrender. As he came to within 150 yards, they threw away their weapons and ran towards him.

'In an instant I was surrounded and hoisted on Italian shoulders,' he recalled. '"*Eviva Germania*!" sounded from a thousand throats. An Italian officer who hesitated to surrender was shot down by his own troops.' Rommel had captured 1,500 men of the 1st Regiment, Salerno Brigade. By the end of the day, his total haul since the start of the offensive was 150 officers, 9,000 men and 150 guns. For this outstanding service he was awarded the Pour le Mérite – the 'Blue Max' – Imperial Germany's highest award for bravery.

By 27 October, the German Fourteenth Army had advanced more than 15 miles, though the Austrian Second Army had managed just 4. With the whole of his north-east front in danger of collapsing. Cadorna at last ordered a withdrawal behind the Tagliamento. Still largely intact, his

Third Army set off in good order; but confusion quickly set in as the withdrawal routes became clogged with military traffic and civilian refugees.

To give the Third Army time to get away, the Italian Second Army was told to hold an intermediate line on the River Torre until the 29th. But its demoralized units could provide only piecemeal resistance and German troops were across the Torre and driving towards Cadorna's headquarters at Udine by the early hours of the 27th. Split in two, the Second Army was rapidly disintegrating.

Now began the race for the Tagliamento, with the victorious Austro-German troops hampered only by fatigue, lengthening lines of communication, and the confusion of various formations cutting across each other's axes of advance. In the event, the race was narrowly won by the Italians. But they had already lost 180,000 troops and 1,500 guns. 'In some sectors,' wrote Falls, 'the Italian leadership had completely broken down and whole divisions had disappeared with their artillery.' A post-war commission of inquiry came to the conclusion that many artillery batteries rode away 'with their officers at their head, deserting their infantry'.

Despite the fact that the original aim of the offensive had now been achieved, the Austrians were keen to press on. General Erich von Ludendorff, the German Chief of Staff, agreed to extend the loan of his troops until they had reached the River Piave. On 31 October, and for the next two days, the Austro-Germans tried without success to cross the Tagliamento. It seemed as if the Italians would hang on, and Cadorna now told the French and British that he would no longer need the twenty divisions he had asked for; the six they had already begun to send would suffice.

But all this changed on the night of 2 November when the German 55th Division secured a bridgehead near Corninho. Two Italian divisions – the 36th and the 63rd – were taken in the flank and thousands of their soldiers captured. In the afternoon of 3 November, having informed his government that the Tagliamento line had collapsed, Cadorna requested permission to retire to the River Piave. At the same time, the Allies were asked for fifteen divisions to shore up the defences there. They eventually agreed to send up to twelve divisions (six British and six French), but made it clear that a change in command was imperative.

On 7 November, having just issued an Order of the Day calling on his men to 'die and not to yield' on the Piave, Cadorna was informed of his dismissal. By noon on the 9th, the day his replacement, General Armando Díaz, took command, the last Italian troops crossed the river and shortly afterwards the remaining bridges were blown.

By now the Austro-German advance was running out of momentum. Despite having a nominal superiority of fifty-three divisions to thirty-three, many were down to one-third of their establishment whereas the Italian formations were close to full strength. Furthermore, the Italians were on a strong natural position and would soon be bolstered by the arrival of five British and four French divisions.

Various attempts were made to force the river after 15 November, and small footholds were gained, but overall the operation was a failure. The Austro-German lines of supply were so strung out, and transport so scarce, that there were not enough boats or bridging materials to make a crossing possible. With winter closing in, the Austrians eventually decided to postpone further offensives until the following spring.

Nevertheless, particularly by the standards of the First World War, Caporetto was a major defeat for the Italians. In just two weeks, the Austro-German armies had forced them back 50 miles and captured 3,000 guns and vast amounts of equipment. They had also inflicted more than 300,000 casualties. That 85 per cent of them were taken prisoner is an indication of how completely Italian morale had collapsed.

The Kaiser's Battle

On 21 March 1918, the Germans made one last attempt to win the First World War when they attacked 50 miles of British front between Arras and the River Oise. By nightfall, they had gained almost 100 square miles of ground. It was one of the most spectacular successes of the war and due, in no small part, to the feeble defence put up by British troops.

The decision to launch the spring offensive had been taken by General Ludendorff, the German Chief of Staff, the previous November. For time was working against Germany: the United States had joined the Allies the previous year and American troops were arriving in France in ever increasing numbers; the Allied blockade was causing severe shortages in Germany, particularly of food and raw materials; her allies, particularly Austria–Hungary, were on their last legs. It was now or never if her ruling elite wanted to avoid the catastrophe of a negotiated peace.

In Germany's favour was the fact that, having seized power in Russia in October 1917, the Bolsheviks had at once made clear their determination to seek peace. This had led to an armistice with Germany in early December 1917 and, after much negotiation, the signing of the Treaty of Brest-Litovsk on 3 March 1918. At last, Germany was released from the burden of fighting a war on two fronts; by March 1918, Ludendorff would have transferred fifty-two divisions from Russia to the Western Front. This, and the fact that Germany had not conducted an offensive in the West since 1916, gave his armies a numerical advantage of 192 divisions to the Allies' 169.

However, a superiority in troops had guaranteed neither side victory before – as Haig knew to his cost from the Somme. Ludendorff hoped to break the stalemate by the use of revolutionary new tactics designed to catch his enemy off guard. Trains would be active all along the front to disguise the point of attack, and assaulting troops would only move into position at night. There would just be a brief preliminary bombardment and the attack would be carried out by small groups probing for the weak spots in the enemy line.

Like Haig at Passchendaele, Ludendorff saw Flanders as the key to the front, specifically the area just south of Ypres. But an attack here would form the second stage of his offensive. The initial assault would go in against a massive 50-mile stretch of the British front from La Fère in the south, near the junction with the French, north almost to Arras. The ground was drier here than in the northern British sector, in Flanders, and the defending forces were known to be weaker. Once a breakthrough had been achieved, the German troops would begin to 'roll up' the British front from the south, while the second attack took place in the north. The bulk of the British Army would either be destroyed piecemeal or driven on to the Channel ports.

In honour of his imperial master, Kaiser Wilhelm II, whose political future depended on the outcome of the spring offensive, Ludendorff named it *Die Kaiserschlact* – 'The Kaiser's Battle'

The sector chosen for the initial attack stretched from the Somme region in the south to Artois in the north. Its lower boundary was the wide and marshy River Oise, its upper limit the River Scarpe near Arras. But between these two river valleys was an extensive ridge that gave the countryside its rolling nature. The rich topsoil was well drained by the chalk beneath, making it perfect for agriculture; wheat and sugar beet had been grown before the war and there were few ditches or hedgerows. Man had added numerous sunken roads and quarries.

For most of the war, this area had been well behind the German lines. But in early 1917, having been pushed out of their strong positions further west during the Battle of the Somme, the Germans moved into this new defensive position from Arras to Soissons. They named it the *Siegfried Stellung* (the Siegfried Line); the Allies called it the Hindenburg Line, after the German Commander-in-Chief, Field Marshal Paul von Hindenburg. During 1917, it was dented by two British offensives: the Battle of Arras left the British holding a small salient at Bullecourt; the tank attack at Cambrai further south put them in possession of an even bigger one known as the Flesquières Salient.

Initially, the British front had only stretched as far south as the German-held town of St Quentin. But in January 1918, at the request of the French, the British took over two further stretches of the line; the new junction between the Allied armies was near the village of Barisis, 5 miles south of the River Oise. The Germans had intended to attack the original junction; now their assault would fall on the British alone.

Field Marshal Crown Prince Rupprecht of Bavaria's army group was occupying the entire stretch of front from which the attack was to be made. But Ludendorff wanted the Kaiser's son, Crown Prince Wilhelm, another army group commander, to be involved and gave him responsibility for the southern third of the attack. Three armies from these two army groups would be involved.

In the north and centre, Rupprecht had von Below's Seventeenth Army and General G. von der Marwitz's Second Army. Both these Prussian generals had impressive records: von Below had served under Ludendorff during the decisive victory over the Russians at Tannenberg in 1914, and more recently had defeated the Italians at Caporetto; von der

Marwitz, a cavalryman, had had many successes on the Western Front, notably the planning and execution of the German counter-attack at Cambrai. Because of this, they were given the most important tasks. Von Below would break through to Bapaume, and von der Marwitz would aim for Péronne; they would then wheel right and, side by side, roll up the British line to the north.

Attacking in the southern sector would be the Eighteenth Army from Crown Prince Wilhelm's army group, under General Oskar von Hutier. He was another of Ludendorff's Eastern Front protégés and had overseen the great German victory over the Russians at Riga in late 1917. His task was to pierce the British lines near St Quentin, advance to Ham and then form the flank guard as the other two armies wheeled north.

Defending the front the Germans were planning to attack were two British armies – the Third and the Fifth. In command of the Third Army, due to face the northern third of the German assault, was a cavalry officer, General the Hon. Sir Julian Byng, who had commanded the Canadian Corps that captured Vimy Ridge during the Battle of Arras in 1917. He had also planned and set in motion the tank attack at Cambrai, but there the early gains were mostly lost when the Germans counter-attacked, and Byng was lucky to stay in his post. His sector stretched from 3 miles north of Arras down to the southern edge of the Flesquières Salient. Haig had wanted him to withdraw from the salient, arguing that it was dangerously exposed, but Byng held his ground. It was a decision that was to cost his men dear.

In charge of the Fifth Army, against which the southern two-thirds of the German attack would fall, was Lieutenant-General Sir Hubert Gough, the man who had commanded the unemployed cavalry of the Reserve Army (later Fifth Army) on the Somme (see p.97). The following year, he had played a central role in the series of futile attacks round Ypres that came to be known as Passchendaele. So great were the casualties that the 'Thruster' of earlier years became known as the 'Butcher'. The Prime Minister, David Lloyd George, had pressed for Gough's removal, but Haig (who had been promoted Field Marshal in December 1916, a month after the Somme battle ended) stood by him; when the British front was extended, Haig sent him to command the new sector because he did not expect any major action there.

Of the four armies that made up the British Expeditionary Force in March 1918, the Fifth was by far the weakest. It had just twelve infantry and three cavalry divisions to defend a front of some 42 miles. Furthermore, the cavalry could only produce a trench-fighting strength of one infantry brigade, giving him the equivalent of just thirteen infantry divisions. The Third Army, in comparison, had fourteen divisions for just 28 miles of front, while the two British armies further to the north were of similar strength. Haig had deliberately concentrated his resources in the centre and north of his line because he feared a German breakthrough in either place. One would cut his armies in two, the other would sever his communications with the Channel ports.

In the unlikely event of the Fifth Army being attacked, Haig had instructed Gough to stand firm for as long as he could, and only if severely pressed to conduct a fighting withdrawal to his final line of defences. The French had agreed to provide six divisions in the event of such an emergency.

As almost four years of war had demonstrated that no single line of trenches was impregnable if attacked with sufficient force, the British now based their defence system on the principle of 'elasticity'. The front line of trenches, now classified the 'Forward Zone', was regarded as an outpost line, held in sufficient strength for troops to inflict maximum casualties on any attackers before withdrawing.

The main defences, forming what was known as the 'Battle Zone', were sited in the best tactical position and far enough back to avoid being destroyed in the initial assault. These defences were the final defences and had to be held at all costs. On average, the two zones were between 2,000–3,000 yards apart, with the Battle Zone roughly the same distance in depth. It was only partially manned, however; most of its defenders were able to live in billets near by, ready to take up their positions in the event of an attack.

Further back were the reserves, ready to counter-attack any section of the line threatened. A 'Corps Line', a third system of defences, was planned, but was still far from completion when the blow fell.

Needless to say, the defences were least satisfactory in the Fifth Army's sector, where the British had arrived late, the front was too long, and there were too few men. Though well protected by barbed wire, the forward trench was not even continuous. Instead, scattered outposts, sometimes manned by as little as a section (the smallest sub-division of a British infantry battalion, commanded by an NCO), watched the front, hoping to cover the empty ground on either side with machine-guns. Just behind the front line, to the rear of the Forward Zone, was a series of redoubts, all-round defensive positions sited on natural features dominating the surrounding land. They were usually too far apart to be able to give effective supporting fire. A long way from completion, particularly in the extreme south of the sector, the Battle Zone was not much of an improvement, while the Corps Line was almost non-existent.

Most Fifth Army brigade sectors had one battalion in the Forward Zone, one in the Battle Zone and one in reserve. Half of the foremost battalion would be in the outposts, half in the redoubts. When the battle began, therefore, many of these troops would be cut off from battalion headquarters and forced to rely on their own initiative.

'I was very worried about the scanty way the front line was held,' recalled a captain in the 2nd Leinster Regiment. 'The trenches were poor and shallow and the wiring in front consisted of a single strand of barbed wire held by screw-iron stakes here and there and, in stretches, this was on the ground forming no defence at all. All we had in the company front-line trench was a pair of sentries about every hundred yards on a long stretch of front.'

Needless to say, soldiers prefer to fight in line, with comrades on either side, and there is no doubt that these isolated defenders assumed that they were considered to be expendable. It is hardly surprising, then, that their performance in battle was not all it could have been.

Twenty-five British divisions of the Third and Fifth Armies would be involved on the first day of the battle. Three were Regular, six were Territorial, fifteen were New Army and one – the 63rd – was a naval division, composed of Royal Navy and Royal Marine battalions. The New Army formations, in particular, were under-trained and poorly led. It was Gough's misfortune to have most of the New Army men and none of the Regulars.

During the night of 17 March 1918 – St Patrick's Day – two German soldiers left their trench just south of St Quentin, walked across no man's land and surrendered to Ulstermen of the 36th Division. They were from Alsace, a French province ceded to Germany in 1871, and were not prepared to die for a Fatherland they did not recognize. Under interrogation, they revealed that they were from a mining company whose duty it was to blow away the British barbed wire in preparation for a full-scale attack. It would go in after a six-hour bombardment in which prussic acid gas shells would be fired. This information found its way back to Gough's headquarters, and from there to Haig's General Headquarters in Montreuil. It was added to a plethora of intelligence – including air reconnaissance reports – that confirmed an attack was imminent.

By the morning of 19 March, Haig's intelligence staff had a pretty clear picture of exactly what was about to happen. They knew that the German rear areas were full of troops, that vast quantities of ammunition were being dumped in the forward areas, that artillery pieces had appeared close behind the front line, that a new German gas mask had been issued, and that no great number of tanks would be involved. They concluded that the British Third and Fifth Armies would be attacked on the 20th or 21st after a short bombardment using a high proportion of gas shells.

Warning was duly given to the formations involved, and artillery batteries were ordered to go over to full 'counter-preparation fire' at night, but it was not always heeded by the units on the ground. There had been a similar warning earlier that month and nothing had come of it. In one Fifth Army battalion, seventy-two men, including the commanding officer, went on leave as late as 20 March.

Only General Gough seemed to be taking the threat of an attack seriously. During the evening of 19 March, he requested permission to move the two GHQ reserve divisions – the 20th and the 50th – closer to his front line. Lieutenant-General the Hon. Sir Herbert Lawrence, Haig's Chief of Staff, refused to give it and a despondent Gough closed the conversation with the words: 'I shall fight them in my Battle Zone as long as we can hold them there. Good night. Good night.'

That same evening, thousands of Germans due to take part in the initial assault marched into their forward zone and took cover in a number of shelters – dugouts, cellars and special assembly areas that had been dug behind the front line. They remained hidden there, within easy range of the British artillery, for the whole of the next day. Once under the cover of darkness, they moved forward to their jumping-off points.

At exactly 4.40am, almost half the German artillery on the Western Front – 6,473 guns (more than a third of them 5.9-inch 'heavies') – opened up over a 50-mile front. Also in action were 3,532 mortars, capable of firing bombs weighing up to 100 kilos. During the 5-hour bombardment the guns fired 1,160,000 shells. The British, by comparison, had fired 1,500,000 shells in seven days before the Somme. It was by far the most concentrated artillery bombardment of the war.

'I was going round inspecting the posts,' wrote a lieutenant in the 22nd Northumberland Fusiliers, 'and just happened to be standing on the fire-step, with my head just over the parapet, looking out over No Man's Land. Then I saw this colossal flash of light. As far as I could see from left to right was lit up by it . . . I think I just managed to hear the gunfire itself before the explosions as the shells arrived all around us.'

As the storm of steel descended, the British infantry in the front line took what cover they could. Some were in dugouts, safe from all but a direct hit by a heavy shell, though the walls shook and dry earth kept trickling through the joints in the roof timbers. But most were in open trenches, huddled into corners and crammed into scrape holes in the wall. The shells exploding all around them were like 'millions of saucepans all boiling together'. Then they realized to their horror that many contained gas – chlorine, phosgene and lachrymatory (tear) gas. Even those lucky enough to get their gas masks on in time were severely debilitated: breathing was difficult through the primitive filters and visibility restricted by the poor eyepieces.

At 9.40am, the bombardment ceased and the storm troops of 32 German divisions began to advance. They were greatly assisted by a dense fog. 'The infantry commander shouts, "*Drauf!*" and we rush forward,' recalled one German pioneer. 'But where is the expected enemy fire! There is hardly any . . . Then we reached the barbed wire, our objective. But there is nothing for us to do. The wire is completely destroyed. There wasn't really any trench left, just craters and craters . . . Only a few of the enemy had survived the storm; some were wounded. They stood with hands up.'

This account was typical. The German artillery and mortars had done their work well. Most of the wire had been blown away and the front-line trenches were largely cratered ruins. However, in many places the British had deliberately evacuated the forward trench, often without orders. The few who were left were either dead, wounded or so shocked and gassed as to be incapable of offering any serious resistance. 'It would not be an exaggeration,' wrote Martin Middlebrook, author of *The Kaiser's Battle*, 'to say that nine-tenths of the British front-line trench or outposts fell without much of a fight.'

But the Germans had concentrated on punching holes in certain sectors, leaving isolated pockets to be dealt with later. Some fought bravely, others quickly surrendered. Only a few got back to the next line of trenches. Within an hour, the Germans were in possession of all 50 miles of the British front line from just south of Arras to the River Oise. The only exceptions were the Flesquières Salient, which had not been attacked, and a few scattered outposts.

The next phase of the battle was fought in the main positions of the Forward Zone. In general, the Third Army had support and reserve trenches, while the Fifth Army had to make do with redoubts and smaller supporting posts. Nevertheless, these positions were considerably stronger – with a greater variety of weapons and better artillery support – than those of the front line, and their defenders should have put up a better fight. But before long they too began to crumble, particularly in the south of the Fifth Army's front, where the defenders were weakest and the fog was thickest.

At the village of Fayet, just north of St Quentin, part of the 2/8th Worcestershire Regiment, a Territorial battalion, was attacked by the 109th Leib Grenadier Regiment. 'We followed up the creeping barrage quickly,' wrote a German corporal, 'but as soon as we appeared the British threw away their weapons and surrendered. There was really no fight for Fayet.'

Judging by its casualties, the 7th Sherwood Foresters, also a Territorial unit, which had suffered heavily at Gommecourt on the first day of the Somme, must have put up a stiffer resistance. Even so, its whole Forward Zone was overwhelmed in less than an hour, with just 2 wounded officers and 12 men escaping back to the Battle Zone. Of the remainder, 12 officers and 470 men, many wounded, were taken prisoner, while another 12 officers and 159 men were killed, the highest proportion of fatalities for any battalion on this day.

By 11am, only a handful of redoubts were still holding out in the main positions of the British Forward Zone. To all intents and purposes, for 11 miles north of the Flesquières Salient and 38 miles south, the British front-line defences had ceased to exist. No battle on the Western Front had seen such startling success since the trench lines were established there in late 1914.

One of the few islands of resistance was Epéhy, in the north of the Fifth Army sector, which was held by no less than three battalions of the Leicestershire Regiment. There the divisional commander had sensibly ordered the front-line trench to be largely evacuated before dawn, leaving the men concentrated in the formidable redoubt behind. They were further assisted when the fog began to lift, revealing large bodies of German infantry out in the open. 'The Lewis guns [light machine-guns] got busy and the enemy groups scattered,' recalled Lance-Corporal North of the 7th Leicesters. 'They had very little cover and no chance of survival.'

But such stands were of little use if the Germans were able to penetrate on either side of these strongpoints. 'About half a mile to our right,' wrote North, 'we could see the Germans moving forward in single file and many were already well behind us. It was not yet midday. Jerry was moving as if there was no opposition and we reckoned we were in real trouble on the flank.'

In the same boat were the other redoubts still holding out in the Forward Zone. None would survive beyond the evening of 22 March. Most were forced to surrender; a couple, including Epéhy, were able to retire.

So quickly had some German storm troopers penetrated the forward defences on 21 March that they were able to engage and breach part of the Battle Zone within twenty minutes of the opening of their attack – before the fog had a chance to lift. But given the zone's depth of between 2,000 and 3,000 yards, a successful German attack against the front of the Battle Zone and the capture of positions within it did not guarantee complete success. For there were still more defences behind, including the rear trench, known as the 'Brown Line'.

Nevertheless, by late afternoon the Germans had made spectacular gains. Of the four British divisions that had opposed the attack in the Third Army sector above the Flesquières Salient, two, and a part of a third – a brigade of the 34th, and the whole of the 51st (Highland) and 59th (North Midland) Divisions – had been pushed back to the rear edge of their Battle Zones. The 6th Division, situated between the 51st and the 59th, was still holding out inside its Battle Zone, but with its flanks exposed it was forced to withdraw that night.

The salient itself, held by four divisions, had still not been attacked but would almost certainly have to be given up to prevent its defenders from being outflanked.

From the salient down to the Somme Canal in front of St Quentin, a front of 19 miles, only a single division, the 16th (Irish), had been driven to the rear of its Battle Zone. The other five – the 21st, 66th (East Lancs), 24th, 61st (South

Midland) and 30th – along with the South African Brigade, were either still holding the front of their Battle Zones or had halted the Germans within them.

It was south of here, on the stretch of front taken over from the French only eight weeks earlier, that the real disaster had occurred. All three divisions – the 36th (Ulster), 14th (Light) and 18th (Eastern) – and one brigade of the 58th (London) Division, had been all but driven from their Battle Zones and large gaps existed between the few battalions that remained.

A major factor was the tendency of British troops to surrender or fall back without sufficient cause. 'The effect of such retreats,' wrote Middlebrook, 'was that, right from the beginning of the battle, there developed an attitude of uncertainty about flanks, a tendency for men in good defensive positions to look over their shoulders and wonder if they too ought not to be moving back.'

By midnight, the Germans had taken 98½ square miles of ground, all but 19 from the hapless Fifth Army. On the Somme in 1916, the British and French had taken 140 days of hard fighting to capture the same amount of ground. In addition, during the night of 21 March, the British would voluntarily withdraw from a further 40 square miles of ground to prevent many units from being encircled.

German casualties for the day's fighting were 10,900 killed, 28,800 wounded and just 300 taken prisoner. British losses were 7,500 killed, 10,000 wounded, and 21,000 taken prisoner. The total casualties for both sides – around 78,000 – make it the costliest single day's fighting of the war on the Western Front, although the Somme saw more killed. That the total of casualties was shared roughly equally by the two sides is indicative of the feeble defence put up by many of the British troops. For attackers invariably suffer much higher casualties than their opponents: on the first day of the Somme, for example, the British lost some 57,000 men to the Germans' 8,000.

Over the coming days, Ludendorff fed more and more troops into the southern sector where Crown Prince William had won the most spectacular first-day gains. This, in turn, led to more huge advances and a continual hammering of Gough's Fifth Army; but the German push failed to achieve a decisive breakthrough. Instead, it took the pressure off British forces to the north and forced the French to face the threat on their flank.

The turning point came on 26 March, at an Allied conference in Doullens, when Haig agreed to place himself and his troops under General Ferdinand Foch, the French commander. Foch eventually deployed fourteen French divisions to assist the British, using them to counter-attack rather than simply to plug holes in the line. On 5 April, the Ludendorff offensive ran out of steam just 10 miles from Amiens. The Germans had advanced 40 miles and captured 1,000 square miles of ground; but they were still 50 miles from the sea and the same distance from Paris.

Like the Battle of the Bulge in the Second World War, the Kaiser's Battle represented a last throw of the dice for the German military. Anything short of complete victory meant ultimate defeat. That duly came about with the unconditional surrender of 11 November. Nevertheless, 21 March 1918 remains one of the darkest days in British military history. The German tactics, inadequate defences and fog all played their part. But the scale of the disaster can only be put down to the poor fighting performance of the British troops.

'Morale wasn't high,' wrote a lieutenant in the 2/5th Manchester Regiment. 'If we'd had the guts we'd had eighteen months earlier, the Germans would never have knocked a hole in the line as they did that time. I heard later that most of the battalion became prisoners without putting up much of a fight . . . After the war I met some of the officers who had been taken prisoner and had the feeling that they were all rather ashamed of what had happened.'

Anual

On 22 July 1921, at Anual in north-east Morocco, a large Spanish army was routed by a force of Rif tribesmen a fraction of its size. It was one of the worst defeats of a modern army by untrained troops in the history of war, and due in no small part to the abject defeatism shown by the Spanish soldiers and their officers.

At the turn of the twentieth century, both France and Spain had been casting covetous eyes towards Morocco, one of the few African states to have avoided colonization. But France had one major advantage: she was still a great power, whereas Spain had long since relinquished that status. This fact was confirmed by the Anglo-French Entente of 1904, which gave France a free hand in Morocco in return for an undertaking not to interfere with British plans in Egypt. Spain's agreement was secured by a French promise that north Morocco would be treated as a sphere of Spanish influence.

When Morocco became a French protectorate in 1912, this arrangement was confirmed. In effect, Spain was given administrative control over a strip of northern Morocco, about one-tenth of the whole country. It included the Mediterranean ports of Cueta and Melilla, which Spain had held for centuries, and the iron mines of the Rif Mountains. Inevitably, Spanish capital became interested in exploiting these natural resources. But first the rugged and unmapped interior, peopled by fierce, independent tribes, had to be pacified.

The man given the task of extending Spanish rule into the central Rif was Major-General Manuel Silvestre, a lady's man and favourite of King Alfonso XIII, and whose soldierly qualities included his having been wounded no fewer than sixteen times during the Spanish-American War of 1898. Having set out from Melilla with an army of 20,000 Spaniards and 5,000 native *Regulares* in late 1919, he proceeded to conquer more territory in 18 months than had been won in the previous 12 years.

By May 1921, he had pushed his forward posts more than 80 miles west of Melilla. They stretched from Sidi Dris on the coast, over the mountains, through Anual, and south to Zoco el Telata, 35 miles inland. The bulk of his army, about 14,000 men in all, was encamped around the fort at Anual, set in a lush valley surrounded by high ground. The remaining troops were manning a chain of forts and blockhouses that stretched back to Melilla.

Part of Silvestre's early sucess can be put down to the fact that a series of poor harvests had forced many Rif tribesmen to seek work in Algeria. Now they returned and swelled the ranks of those tribes eager to resist Silvestre's advance. None was more implacable than the Beni Urriaguel, led by Abd el-Krim.

Born in the tribal stronghold of Ajdir in 1882, the son of a schoolteacher, el-Krim had been educated at a Spanish school in Melilla and later became the editor of *El Telegramma del Rif*. He had also served for a time in the Bureau of Native Affairs, and there learnt about the Spanish plans to exploit the mineral potential of the Rif. Foolish enough to suggest that Spain should come to some arrangement with a new Rifian state, he was imprisoned in August 1917.

On his release in late 1918, he returned to his tribal lands to raise a force that could defeat the Spanish. This was no easy task, given the fact that the Rifian tribes had no history of fighting together as a unit; they were more used to fighting each other. Nevertheless, by the spring of 1921, he had managed to weld together a potent fighting machine. While most of el-Krim's warriors were from the Beni Urriaguel, the neighbouring tribes were also represented. 'The Rifian warriors,' wrote David Woolman, author of *Rebels in the Rif*, 'were generally tough, valiant, resourceful fighters, and superb shots, whose endurance gave them an amazing mobility. They would walk as much as 30 miles a night over mountain trails and yet be perfectly fit to attack at daybreak. Each man carried a gun, a long, straight dagger, cartridges, bread and fruit. He slept wherever he happened to be, pulling the hood of his *jellaba* [cloak] over his face.'

The Muslims Rifians also had the advantage of a cause. They would be fighting a *jihad* – a holy war – against the infidel invaders; the Spanish had no such motivation. Far from it – their conscript troops were of a quality among the lowest in Europe, with more than 80 per cent illiterate. They were badly trained and poorly equipped, many with rifles more than twenty years old. An investigation of one unit found that nineteen out of thirty guns were in such bad condition that they could not fire a bullet more than a hundred yards. Pay was less than a third of what the tribesmen got for road-building, and the ordinary soldiers in Morocco were forced to subsist on coffee, beans, rice and bread. Barracks and hospitals were filthy, and losses from malaria high. Little wonder that they would do anything to avoid active service, eating tobacco to stimulate jaundice or applying nettles to minor wounds to make them fester.

Their officers did little to alleviate these hardships. Incompetent and undisciplined, many sold army stores to supplement their meagre pay and spent much of the time away from their garrisons, 'gambling and whoring' in Melilla. Hardly surprising, then, that the morale of the Spanish Army in Morocco was at rock bottom.

Silvestre seemed blissfully unaware of his weakness. On being told of el-Krim's warning that any attempt by the

Spanish to cross the Amekram River would be resisted, he remarked: 'I'm not going to take seriously the threats of a littler Berber Caid whom I had at my mercy a short while ago. His insolence merits a new punishment.'

Having built up his supplies, Silvestre intended to cross the river and push further west through the defile beneath Mount Abaran. But first, to secure the passage of the defile and guard the entrance of the valley, he constructed a fort on Abaran and manned it with 500 soldiers. It was this act that provoked el-Krim to attack. During the night of 31 May, he and a small band of men scaled the peak and took the fort by storm.

'The . . . Rifis poured over the parapet,' wrote Rupert Furneaux, author of *Abdel Krim*, 'rifle in one hand, long curved knife in the other. Where possible they used their knives; ammunition was precious. They rushed on the Spaniards, shooting, hacking and stabbing. From all around rose the shrieks of the wounded, the groans of the dying.'

It was all over in thirty minutes. When the Spaniards at Anual realized what was going on, they dispatched a relief force. But it was driven back by the fire from the captured Spanish cannons.

On hearing the news, General Berenguer, the High Commissioner, sailed from Cueta and met Silvestre on board his ship off Sidi Dris on 5 June. There Silvestre tried to convince him that the loss of the fort was a minor setback. But Berenguer feared a widespread revolt and told Silvestre not to advance any further into the Rif. A violent argument ensued, during which Silvestre attempted to throttle the High Commissioner and had to be restrained by his staff officers. A breathless Berenguer then left, under the impression that his instructions would be carried out.

El-Krim, meanwhile, was busy trying to encourage other Rif tribes to join the revolt. 'Oh, Muslims,' he told them, 'we have wanted to make peace with Spain, but Spain does not want it. She only wants to occupy our lands in order to take our property and our women, and to make us abandon our religion. Do not expect anything good of Spain . . . The Koran says, "Who dies in Holy War goes to glory."'

The result was that the neighbouring tribes rose, swelling his *harka* to more than 3,000 men. On 15 July, these warriors took up positions on a stretch of high ground at the south-western end of the Anual valley. Silvestre responded by sending troops to occupy Igueriben, a low hill 4 miles to the south of Anual. It was an unfortunate choice. As it was also 3 miles from the nearest water supply, it would need to be revictualled over an area of rocky terrain through which ran a deep gully.

On 17 July, el-Krim's men dug in on the high ground between the two Spanish outposts. The following day, Major Benitez, the commander at Igueriben, heliographed that he was short of water. Two attempts to reach the stream in the gully had failed, driven back by tribesmen who had surrounded the outpost. A relief force would be sent the following day, he was told.

At dawn on the 19th, three columns set off from Anual, each of 1,000 men. The entrenched Rifians allowed the Spanish to get within 200 yards of their positions before they opened fire. 'The bullets flew like grain,' wrote el-Krim in his memoirs. At first the Spaniards wavered, then they came on again. But the storm of fire was too great and they eventually withdrew, having lost 152 men in two hours.

That afternoon Silvestre, who had arrived at Anual from Sidi Dris, radioed to Berenguer for reinforcements. Replying that there were none available, Berenguer advised him to remain on the defensive. The Igueriben garrison made two attempts to reach the stream in the gully on the 20th; both were unsuccessful. Desperate with thirst, its men had been reduced to drinking the juice from tomato tins, then vinegar, cologne, ink, and, finally sweetened urine.

On 21 July, Silvestre made one last attempt to reach Igueriben with a force of 3,000 men. But again the Rifian fire was too hot. When the smoke cleared the Spaniards could be seen streaming back towards Anual, many having thrown away their weapons. Watching the débâcle from the parapet of the fort at Anual, Silvestre seemed to lose his head. First he sent a heliograph message ordering Benitez to surrender the garrison; but Benitez refused, saying he would rather die. Then he sent a final message: abandon the post and fall back on Anual. In an attempt to do so, Benitez and most of his garrison were hacked to death. Of the twenty-five men who did make it back to Anual, sixteen died soon afterwards of dehydration and exhaustion.

Silvestre now began to panic. The camp at Anual was in a bad location, overlooked by surrounding hills, and his men were displaying little stomach for a fight (the fact that they had only been issued with 40 rounds a man did not help). During the afternoon of the 21st, as the Rifians turned their attention towards Anual, Silvestre called his officers to a council of war. Ammunition was short and most were of the opinion that to hold on was pointless. Silvestre agreed and sent a final telegram to Melilla at 4.55am on 22 June, announcing his intention to withdraw to Ben Tieb. He then ordered a general retreat.

But the shock of such an order seemed to have snapped the already low morale of the conscript troops. Instead of retiring in good order, they broke and ran, throwing away their equipment and weapons as they went. Hundreds were shot and hacked to death by the waiting Rifians. Silvestre, who at the last minute decided not to abandon Anual, is said to have shouted after his panicking troops: 'Run, run, the bogeyman is coming.'

Eventually the fort was taken and many of the inhabitants massacred, among them Silvestre; 800 others were taken prisoner. Rifian legend has it that he was chopped to pieces and scattered to the wind. Another version is that el-Krim personally cut off his head and had it carried through the mountains as proof of the Rifian victory. Whichever account is true, his body was never found.

'Afraid of what?'

Of the 800 Spaniards who were taken prisoner by el-Krim after the Battle of Anual, at least five were young women. Unlike the men, who were put to work road-building ('dejected, despairing and fearful of their fate'), they were well looked after.

One, by the name of Cipriana, was taken in by el-Krim and put in the care of his wife 'for her protection', though he felt 'she didn't want to be protected'. She helped in the home, looking after his children, and was released at the end of the war and repatriated to Spain. Later returning to Morocco to live with her husband at Alhucemas, she died in 1950. On hearing of el-Krim's escape to Egypt, two years before her death, she wrote to thank him for the kind treatment she had received during her five years of captivity.

Another, called Isabella, was fifteen when she was captured. Paul Mowrer, an American author, met her in Ajdir in 1924 and asked if she was afraid. 'Afraid of what?' she answered, explaining that she was well treated and had nothing to complain about. 'The natives liked Isabella,' wrote Mowrer. 'Her smiles and chatter, so different from the subdued reserve of their own women, were apparently irresistible.' Whenever she travelled by mule, Mowrer noted, there was always some young Rifian ready to lift her on and off the saddle by seizing her around the thighs. 'The more she was able to addle the young fellow's wits, the more she was pleased.'

A blonde woman, known as 'La Rubia', is said to have died a few weeks before the end of the war. The other two women prisoners were Maria, who married a Rifian but was later widowed, and her sister Adriana. Both were eventually repatriated.

Most of the men were not so lucky. When the Spanish government paid £150,000 for their release in January 1923, only 350 were still alive.

As the handful of survivors reached posts to the rear, and news of the catastrophe spread, other tribes joined in the rebellion and the chain of Spanish forts collapsed like dominoes. Many were evacuated long before they were threatened. Those that did try to make a stand were soon overwhelmed by the rebels – and sometimes by their own deserters. Everyone that could – soldiers and civilians – fled towards Melilla, leaving the wounded and sick at the mercy of the bloodthirsty tribesmen.

About three thousand survivors were rallied at Dar Drius by General Navarro, Silvestre's second-in-command. So poor was their physical condition, and so low their morale, that he decided to make for Melilla. Harried by rebels as he fell back along the single dirt road, he refused to abandon the wounded and was forced to take refuge in the old fort at Monte Arruit. But, again, the water supply was outside the perimeter and the defenders were tormented by thirst. A lack of medical supplies meant that 167 men died of gangrene alone. Aircraft from Melilla tried to drop supplies but most of them were collected by the Rifians.

On 9 August, Monte Arruit, the last surviving fort outside Melilla, fell to the tribesmen. Navarro had agreed to a formal surrender, but as soon the gates were opened the Rifians poured into the fort and slaughtered most of the defenders. Navarro and 600 men were spared.

Although only 20 miles away, the garrison at Melilla made no attempt to relieve Monte Arruit. Just 1,800 strong, and of doubtful fighting quality, it could not have accomplished much. In any case, the heights of Gurugú, which dominated the city, were swarming with rebels. The Rifians did eventually penetrate the outskirts of the city, causing the 40,000 inhabitants to flee in terror on to boats or inside the old fortified town. But they went no further. Without cannon, an assault against the old fortress would have been costly for the attackers. In any case, the harvest was due and the tribesmen were anxious to return to their land with their booty. As one grizzled warrior put it, the Rifians were simply bored with cutting throats. By the middle of August, el-Krim's army had melted into the interior.

On learning of the disaster at Anual, Berenguer told the press: 'All is lost, including honour.' This was no exaggeration. The final report to the Cortes, the Spanish Parliament, listed 13,000 soldiers killed; other sources put the figure of the dead as high as 19,000. Among the arms and equipment lost were 20,000 rifles, 400 machine-guns and 129 artillery pieces, plus huge stores of ammunition and canned food. In addition, the Spanish forfeited all their investment in north-eastern Morocco: railways, mines, agricultural equipment, schools, military buildings. Twelve years of hard work had been destroyed in just twenty days by a band of tribesmen. Incredibly, the number of Rifians who had attacked Anual was just 3,000.

Following his victory, Abd el-Krim proclaimed the Republic of the Rif. It would last just five years; but it took a combined Franco-Spanish force of 250,000 men to subdue el-Krim. In May 1926, he surrendered to the French and was exiled. Spanish rule in the region continued until the French granted Morocco its independence in 1956.

Like most military disasters, Anual was caused by a number of factors: political pressure, poor leadership, failed strategy. But only the abject performance of the troops, both officers and men, can explain the sheer scale of the catastrophe.

Crete

In May 1941, German airborne forces wrested the strategic island of Crete from a British and Dominion garrison that initially outnumbered them by more than four to one. A reckless gamble, it came off because of the ineptitude of the defenders, particularly the New Zealanders.

The chain of events began in April 1939 when, in response to Italy's invasion of Albania, Britain guaranteed Greek independence. Eighteen months later, with British troops in the Middle East occupied by Graziani's incursion into Egypt (see p.188), the Italians attacked Greece. Churchill was only in a position to send RAF bombers, but they were hardly needed as the Greeks drove the woeful Italian army back into Albania.

In January 1941, however, Churchill received confirmation from decrypted enemy signals (a source later known as Ultra) that a build-up of German troops in Romania posed a serious threat to Greece. General Wavell, then the British Commander-in-Chief in the Middle East, was instructed to offer the Greeks the support of three divisions. The proposal was rejected by General Metaxas, the Greek Prime Minister, on the grounds that the force was large enough to give the Germans an excuse to invade, but too small to stop them.

However, Metaxas died from throat cancer on 29 January, and his replacement, Koryzis, lacked his political insight. The new government quickly made it known that *any* British assistance would be welcome. Churchill was delighted, and it was eventually agreed that 'W Force' – three infantry divisions, one infantry brigade and one armoured brigade – would be sent to defend Greece. In the event, only two divisions (the New Zealand and 6th Australian) and one armoured brigade (1 British) were dispatched. At the last moment, General Wavell kept the other troops back in North Africa to meet the new threat posed by Rommel's recently arrived Afrika Korps. It was just as well.

On 6 April, the Germans launched simultaneous invasions of Greece and Yugoslavia. Defending the 'Metaxas Line', from the River Nestos in Thrace and along the Bulgarian border as far as Yugoslavia, the Greeks fought tenaciously. But within three days, German motorized troops had outflanked these defences by piercing the junction of the Greek and Yugoslav armies and advancing down the Vardar valley to Salonika. The Greek Second Army in Thrace was forced to surrender; resistance in Yugoslavia collapsed.

Instead of continuing their advance from Salonika to Mount Olympus, where much of W Force was in position, the Germans now exploited the collapse of resistance in Yugoslavia by a second thrust through the Monastir Gap and down the west coast of Greece. This cut off the Greek divisions facing the Italians in Albania and threatened to outflank W Force. On 18 April, with defeat inevitable, Prime Minister Koryzis shot himself. Two days later, Wavell took the decision to evacuate W Force to Crete, leaving the Germans to overrun Greece.

As early as October 1940, General Halder, Hitler's Chief of Staff, had suggested that 'mastery of the Eastern Mediterranean was dependent on the capture of Crete, and that this could best be achieved by an air landing'. With Greece about to fall, Lieutenant-General Kurt Student, commander of the

XI Air Corps, put forward just such a plan – codenamed Operation Merkur (Mercury) – to his superior, Goering, and a meeting was arranged with Hitler.

Student's proposal was to use Crete and Cyprus as stepping stones across the Mediterranean. The former island, he explained, would form an ideal base from which to launch a paratroop assault on the Suez Canal to coincide with Rommel's arrival at the outskirts of Alexandria. Would not Malta be a more suitable target than Crete? queried Hitler. No, replied Student, its small size and regular shape meant that its garrison could deploy rapidly to counter-attack all possible dropping zones; Crete, on the other hand, was long and narrow, with bad internal communications. Hitler was not convinced, and predicted heavy casualties.

A few days later, however, Goering secured his approval. According to Martin van Creveld, author of *Hitler's Strategy*, 'far from being part of any coherent strategy', the operation was 'little more than a sop to Goering, whose air force was destined to play a subordinate role in the coming Russian campaign'. Officially, in Führer Directive No. 28 of 25 April, the operation was 'to be prepared in order to have a base for conducting the air war against England in the Eastern Mediterranean.'

Student's plan, outlined to his senior officers in Athens on 15 May, was to divide his forces by targeting all three airfields on the north coast of Crete. Each were near ports where supplies and reinforcements could be landed if the operation was successful. The initial assault would be carried out by the 8,000 men of the élite 7th Parachute Division, the only one the Germans had. Its largest unit, the Storm Regiment, would drop on and around Maleme airfield in the west of the island; the 3rd Regiment and the Engineer Battalion would land nearby in the Ayia valley, south-west of Canea, to intercept any Allied reserves; most of the 2nd Regiment would drop on

Rethymnon airfield, 40 miles to the east; the 1st Regiment would take the capital, Heraklion, and its airfield, a further 50 miles to the east.

Once the airfields were secured, on Day 1 or Day 2, transport planes would fly in the infantry of the 5th Mountain Division (originally, this was to have been the 22nd Airlanding Division). Motorcycle troops, artillery and engineers would follow. About 14,000 men would be flown in in this way. Also on Day 2, two groups of light ships – seven small freighters and sixty-three motor-assisted caiques – would land two battalions of mountain troops, anti-aircraft batteries and supplies. Light tanks and motor transport would arrive once 'shipping communications between the mainland [of Greece] and Crete' had been secured. This seaborne element of the plan was added at Hitler's insistence.

The main weakness of the plan was that it dispersed the limited resources of the parachute division rather than concentrated them. This made the capture of any one airfield – the prerequisite to success – more difficult. Student was prepared to take this risk because his intelligence summaries had seriously misled him as to the exact number of enemy troops on the island. One, on the eve of the battle, 19 May, stated that the British garrison was just 5,000 strong, with only 400 men at Heraklion and none at Rethymnon. All the New Zealanders and Australians had proceeded to Egypt, it claimed, while no Greek troops were on Crete. In fact, with the British garrison of one infantry brigade (14 Brigade) swelled by the survivors of W Force, the island was now held by more than 42,000 servicemen, including 9,000 Greeks.

Student's intelligence staff were also under the illusion that the Cretans would welcome the German troops. They should have read the general briefing document prepared for the invasion of Greece. 'The Cretans,' it stated, 'are considered intelligent, hot-blooded, valorous, excitable as well as obstinate and difficult to govern . . . In case of invasion account must be taken of obstinate resistance by the civilian population.'

Fortunately for the Germans, the defenders of Crete proved to be even more incompetent than Student's intelligence officers. Since their arrival the previous November, the British and other forces had had no less than seven commanders. The seventh was Major-General Bernard Freyberg, VC, until then commander of the New Zealand Division, who was appointed to command in Crete at Churchill's insistence on 30 April, the day after his arrival on the island. A great favourite of the British Prime Minister's, famed for his bold exploits during the First World War (he won the VC during the Battle of the Somme) – Churchill once asked him to strip at a country-house weekend so that he could count his twenty-seven wounds – Freyberg seemed the ideal choice. In fact, other than a talent for training troops, he had little to recommend him as a general, for he was obstinate, muddled in his thinking, and reluctant to criticize or sack incompetent subordinates.

On 30 April, Wavell flew to Crete to inform a reluctant Freyberg of his appointment. He brought with him a signal that he had received from Churchill two days earlier: 'It seems clear from our information that a heavy airborne attack by German troops and bombers will soon be made on Crete. Let me know what forces you have in the island and what your plans are. It ought to be a fine opportunity for killing parachute troops. The island must be stubbornly defended.'

Wavell added that the 'scale of attack envisaged was five to six thousand airborne troops, plus a possible seaborne attack'. The 'primary objectives' were 'considered to be Heraklion and Maleme aerodromes'.

Despite being armed with this remarkably accurate information, Freyberg quickly became despondent. Crete, with its mountainous southern coast, could only be resupplied through its northern ports, which were vulnerable to attack by German planes based in Greece. Strong air cover was the solution, but the RAF complement on the island was just six Hurricane fighters and seventeen 'obsolete aircraft'. On 31 April, Freyberg signalled Wavell: 'Forces at my disposal are totally inadequate to meet attack envisaged. Unless fighter aircraft are greatly increased and naval forces made available to deal with seaborne attack I cannot hope to hold out with land forces alone, which as a result of campaign in Greece are now devoid of any artillery, having insufficient tools for digging, very little transport, and inadequate war reserves of equipment and ammunition.'

Four days later, having been informed by Wavell that the Royal Navy would give every support, Freyberg's pessimistic mood had lifted. 'Cannot understand nervousness,' he signalled to London, 'am not in the least anxious about airborne attack; have made my dispositions and feel can cope adequately with the troops at my disposal. Combination of seaborne and airborne attack is different. If that comes before I can get the guns and transport here the situation will be difficult. Even so, provided Navy can help, trust all will be well.'

Both signals show Freyberg's preoccupation with a seaborne follow-up to an airborne attack. The belief had taken root during his meeting with Wavell, and was later confirmed by signals from London based on Ultra intercepts. On 13 May, for example, he was told that the 'invading force . . . will consist of some thirty to thirty-five thousand men, of which some twelve thousand will be the parachute landing contingent, and ten thousand will be transported by sea.'

In fact, military intelligence in London had made the mistake of assuming that the German formation originally earmarked to reinforce the paratroopers, the 22nd Airlanding

Division, was still involved. This led them to conclude that the 5th Mountain Division was an additional formation that would be brought by sea. But even taking this error into account, it is extraordinary that Freyberg failed to grasp that the major element of the attack was going to be from the air. As early as 6 May, a signal from London had given the date of the attack as 17 May and the airlanding strength as two divisions. Even the 13 May signal made it clear that most of the attackers would arrive by air. Yet Freyberg would later admit: 'We for our part were mostly proccupied by seaborne landings, not by the threat of air landings.'

According to Anthony Beevor, author of *Crete*, Freyberg's 'misreading of the threat inevitably produced a damaging compromise both in the disposition of his troops and in his operational orders, which confused priorities'. Six months earlier, Brigadier Tidbury, the first British commander on Crete, had anticipated Operation Merkur by devising a plan to combat airborne assaults on the three north-coast airfields and the Ayia valley south-west of Canea. Freyberg's staff now adapted this plan in the Maleme and Canea sectors to counter an attack from the sea.

The final troop dispositions were as follows: at Heraklion were three regular battalions of 14 Infantry Brigade – the 2nd Black Watch, 2nd Leicestershire Regiment and 2nd York and Lancaster Regiment – reinforced by one Australian and three Greek battalions. They were supported by a medium artillery regiment, six light tanks and two Matildas.

Rethymnon airfield, almost 50 miles to the west, was guarded by two Australian and two Greek battalions, while the town was held by a strong force of Cretan gendarmes. A reserve of two Australian battalions, with most of the motor transport, was stationed a further 20 miles west at Georgioupolis, halfway to Canea. The main reserve – 4 New Zealand Brigade and the 1st Welch Regiment – was near to Freyberg's

'Creforce' Headquarters, in a quarry on high ground to the west of Canea.

But along the crucial 11-mile sector between Canea and Maleme, Freyberg had deployed the rest of the New Zealand Division to cover the coast as well as Maleme airfield and the Ayia valley. This meant their defences had little depth. To compound this error, he placed the divisional reserve, the 20th Battalion, close to Canea rather than to Maleme airfield, the obvious enemy objective. Even more bizarre was the fact that the defences ended on the far edge of the airfield, leaving it vulnerable to an attack from the flank. Aware of this, Brigadier Puttick, who had replaced Freyberg as divisional commander when the latter assumed overall command in Crete, asked for reinforcements to cover the bed of the River Tavronitis to the west. But on 13 May, when the Greek authorities finally agreed to move their 1st Regiment from Kastelli Kissamou, Freyberg cancelled the order; with the invasion imminent, he did not think there would be enough time to complete the march and prepare adequate defences.

Nevertheless, his spirits were high. 'Have completed plan for defence of Crete,' he signalled Wavell on 16 May, 'and have just returned from final tour of defences. I feel greatly encouraged by my visit. Everywhere all ranks are fit and morale is high. All defences have been extended and positions wired as much as possible. We have forty-five field guns placed, with adequate ammunition dumped. Two Infantry tanks are at each aerodrome . . . I do not wish to be over-confident, but I feel that at least we will give excellent account of ourselves.'

On Tuesday, 20 May 1941, the German air raids along the north coast of Crete began earlier than usual, at 6am. The daily 'hate' lasted for one and a half hours – and then the planes disappeared out to sea. Thirty minutes later, there was a second raid by bombers and strafing fighters, concentrating

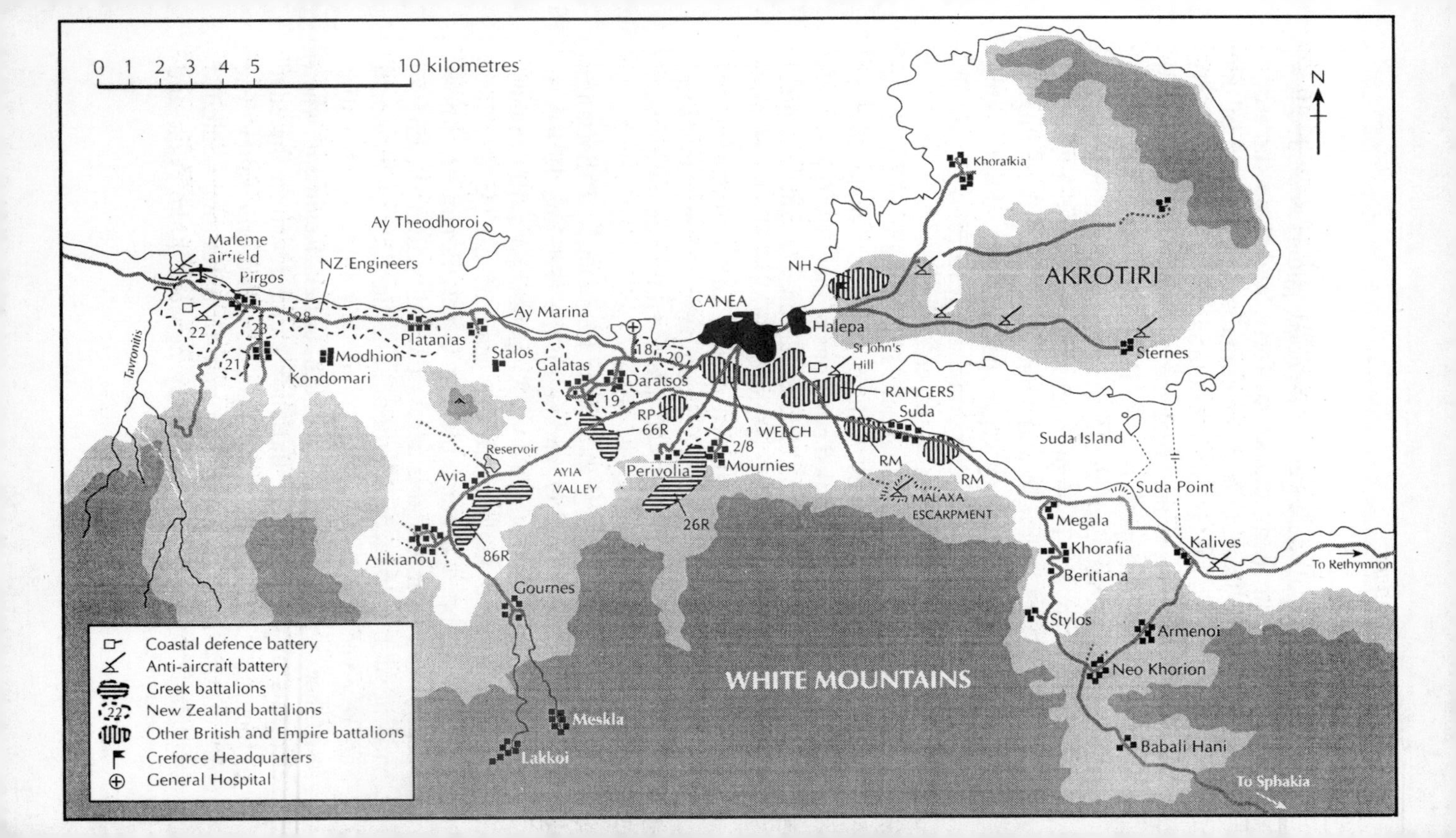
0 1 2 3 4 5 10 kilometres
N
Khorafkia
AKROTIRI
Ay Theodhoroi
Maleme airfield
Pirgos
NZ Engineers
NH
CANEA
Halepa
Ay Marina
Platanias
22
23
28
21
Modhion
Kondomari
Tavronitis
Stalos
18
20
St John's Hill
Sternes
Galatas
Daratsos
19
RANGERS
Suda
RP
66R
1 WELCH
2/8
Reservoir
Suda Island
Ayia
AYIA VALLEY
Perivolia
Mournies
RM
RM
Suda Point
MALAXA ESCARPMENT
26R
Megala
Alikianou
86R
Khorafia
Beritiana
Kalives
To Rethymnon
Gournes
Stylos
Armenoi
Neo Khorion
WHITE MOUNTAINS
Meskla
Lakkoi
Babali Hani
To Sphakia
Coastal defence battery
Anti-aircraft battery
Greek battalions
New Zealand battalions
Other British and Empire battalions
Creforce Headquarters
General Hospital

on the anti-aircraft defences. As it ended, gliders appeared low over Maleme airfield, and about forty landed on the stony riverbed of the Tavronitis. They contained I Battalion of the Storm Regiment, regimental headquarters and part of III Battalion. Minutes later, Junkers transport planes arrived overhead and dropped II and IV Battalions beyond the Tavronitis so that they could use the dead ground of the riverbed as cover. Though badly mauled by fire from D Company of the New Zealanders' 22nd Battalion, dug in along the Tavronitis, they regrouped and set about taking the airfield and Hill 107, the dominant feature to the south, held by C and A Companies respectively.

The rest of the Storm Regiment's III Battalion was not so lucky. Dropping 1¼ miles to the east of the airfield, its men landed on the well-concealed positions of the New Zealander's Engineer Detachment and 23rd Battalion. Caught in a murderous crossfire, many were dead before they hit the ground; the rest were ruthlessly hunted down. The commander, his adjutant, and three out of four company commanders were killed, along with nearly 400 men. Hundreds more were wounded.

A large part of the 3rd Parachute Regiment's III Battalion suffered a similar fate by landing on the 18th and 19th New Zealand Battalions of Colonel Kippenberger's 10 Brigade, grouped around Galatas at the head of the Ayia valley. The rest dropped around the 6th Greek Regiment, whose position ran from Galatas across the valley. But the Greeks had not yet been issued with the ammunition that their colonel had received the day before – and they fled. Much stiffer resistance was offered by the 8th Greek Regiment when elements of the Engineer Battalion came down on the edge of their position beyond Lake Ayia, 3 miles up the valley. The rest of the 3rd Regiment – I and II Battalions – landed virtually unopposed in the valley between Lake Ayia and an abandoned prison, which they quickly occupied. Kippenberger was all for counter-attacking, and requested the release of 20th Battalion, the divisional reserve, to carry out the manoeuvre. But Puttick, backed by Freyberg, vetoed the move because the threat from the sea had priority.

Back at Maleme, Colonel Andrew, commanding the 22nd Battalion, had lost contact with his forward companies and feared the worst. In the early afternoon, he sent up green and white flares, the emergency signal for the neighbouring battalion, the 23rd, to counter-attack. But the flares were never spotted. At 5pm, he finally managed to get through to brigade headquarters, inexplicably sited 4 miles down the coast at the opposite end of the brigade sector. Brigadier Hargest's response to Andrew's urgent request for support was that the 23rd Battalion was itself engaged with paratroopers – which was untrue. Hargest was either misinformed, or particularly conscious of the fact that the 23rd Battalion had been earmarked for coastal defence.

Desperate, Andrew sent forward his two Matilda tanks with the reserve platoon of infantry. One tank turned back after its crew discovered that it had taken the wrong ammunition and could not traverse its turret properly. The other got stuck on the Tavronitis riverbed and had to be abandoned, while the exposed infantry withdrew with several casualties.

At about 6pm, Andrew again contacted Hargest by wireless, and told him that without support he would have to withdraw. 'If you must, you must,' Hargest replied, adding that he was sending two companies. But by 9pm they had not appeared and Andrew was still out of touch with his forward companies. Fearing they had been, or were about to be, overrun – when in fact, despite many casualties, they were still holding out – he ordered a withdrawal to the ridge south-east of Hill 107. Runners were sent to inform the men, but they failed to reach C and D Companies. Meanwhile, one of the companies sent by Hargest got to within 200 yards of C Company's command post at the airfield when it decided that they must have been overrun and turned back.

The other company found Andrew's new position. But after more indecision, Andrew decided that it was too exposed and ordered a further withdrawal back to the 23rd Battalion. During the night, both forward companies discovered that the rest of the battalion had withdrawn, and therefore followed suit. By dawn on 21 May, thanks to this series of blunders, the Germans were in possession of Maleme airfield.

The second wave of attacks were delayed and did not begin until after 4pm on 20 May. At Rethymnon, only two companies of the 2nd Parachute Regiment were dropped in the right place. The rest either fell into the sea or landed right on the Australian positions. One battle group, under Major Kroh, managed to take Hill A to the south-east of the airfield. But the 2/1st Australian Battalion counter-attacked and drove them off it the following morning. Colonel Sturm's detachment of 200 men fared little better. Landing in front of the 2/11th Australians, on Hill B to the west of the airfield, they were easy targets. By morning, most had been either killed or captured. Only Captain Weidemann's battle group, dropping closer to the town of Rethymnon, escaped heavy casualties while still in the air. But they soon came up against the Cretan gendarmerie and armed civilians, and were forced to take refuge in the seaside village of Perivolia.

At Heraklion, II Battalion of the 3rd Parachute Regiment landed between the town and the airfield to the east, an area guarded by no less than three infantry battalions, and was cut to pieces. By nightfall, more than 300 had been killed, 100 wounded and scores taken prisoner. Dropping to the south and east of the town, many members of III Battalion were killed on landing by civilians and Greek soldiers from three ill-equipped regiments. The rest, however, managed to regroup and fight their way into the town from two directions. But confronted by reinforcements from the 2nd Leicesters and 2nd York and Lancasters, they withdrew the way they had come. The 3rd Parachute Regiment's I Battalion suffered the least casualties, dropping 5 miles to the east to take the sparsely defended radio station at Gournes. Over the next couple of days, however, it lost more than 200 men to Cretan guerrilla activity.

Overall, the first twenty-four hours of Operation Mercury had not been a success for the Germans. At a cost of 2,000 paratroopers or glider troops killed, just one objective – Maleme airfield – had been secured, and then only because of the defenders' ineptitude. Even there, the Storm Regiment was so reduced in numbers and so low on ammunition that a determined counter-attack would surely have succeeded. But none came, despite the fact that two New Zealand battalions were stationed less than two miles away.

Puttick eventually gave orders for a counter-attack by the 20th and 28th New Zealander Battalions during the afternoon of 21 May. But Freyberg still feared an assault on Canea from the sea and would not release the 20th until it had been replaced by the Australian reserves at Georgioupolis. Consequently, the attack did not begin until the early hours of 22 May, by which time the German defensive positions had been bolstered by the arrival of a battalion of mountain troops, which had landed in transport planes at Maleme airfield the previous evening.

Nevertheless, with the 21st Battalion also joining the attack, the New Zealanders still had a superiority in numbers and were supported by three tanks. It made little difference. One tank was knocked out by a captured Bofors anti-aircraft gun, another broke down and the commander of the third refused to continue alone. A single company of the 20th Battalion reached the edge of the airfield – but it was the exception.

Most of the New Zealanders did not get within 1¼ miles of their objective; by afternoon the attack had petered out.

Two more mountain battalions landed that day, tipping the balance irrevocably in favour of the Germans. Brigadier Hargest was self-deluded enough to report to divisional headquarters: 'Steady flow of enemy planes landing and taking off. May be trying to take troops off. Investigating.'

Meanwhile, the garrisons at Rethymnon and Heraklion were content to hold their ground in the confident belief that the invasion could not succeed without resupply. Little did they know that events at Maleme had sealed their fate.

Freyberg's fear of a seaborne invasion was never realized. One flotilla bound for Maleme – an Italian destroyer, two steamers and nineteen caiques, carrying a battalion of mountain troops and supplies – was destroyed by a Royal Navy task force 18 miles off the north coast of Crete on the night of 21 May. The second flotilla turned back the following day after encountering a separate task force. But the Royal Navy paid heavily when its ships were caught in daylight by German bombers. One cruiser and two destroyers were sunk, two battleships, two cruisers and several destroyers damaged.

By 25 May, Hargest's brigade had abandoned the Maleme sector and withdrawn behind the new front line from Galatas to the sea. Held by the 18th New Zealand Battalion and a composite force of artillery and transport troops, it was attacked during the afternoon by paratroopers and a mountain battalion. The 18th was the first to crack, quickly followed by the composite force. 'Back, back!' some men cried. 'They're coming through in thousands!'

Kippenberger tried to stem the flow by shouting, 'Stand for New Zealand!' but it made little difference. Only a desperate counter-attack into Galatas prevented a complete rout. That night, the survivors withdrew to positions held by the two

Australian battalions at the end of Ayia valley. Next morning, Freyberg signalled to Wavell that his troops had reached the 'limit of endurance' and 'our position here is hopeless'.

The evacuations began on 28 May. Under the cover of darkness, the troops at Heraklion were picked up by destroyers without incident. But during the voyage to Alexandria, enemy planes sank one destroyer and badly damaged two more, killing about 700 soldiers (a fifth of all those rescued). At Rethymnon, the two Australian battalions never received the order to evacuate and were forced to surrender during the morning of 30 May.

Most of the survivors from the Maleme and Canea sectors were picked up by the Royal Navy at Sphakia on the south coast, having survived a nightmare march across the mountains of central Crete. But when heavy naval losses brought the evacuations to an end, in the early hours of 1 June, 5,000 troops were still waiting to be rescued.

The total British and Dominion losses were fearful: 1,700 killed, the same number wounded and more than 12,000 captured. A further 1,800 members of the Royal Navy lost their lives, while 3 cruisers and 6 destroyers were sunk, and 13 ships badly damaged, including the only aircraft carrier in the Mediterranean. German casualties, though heavy, were minor in comparison: 4,000 killed (three-quarters of them paratroopers) and 2,500 wounded.

Yet for most of the first two days of the battle no more than 8,000 attackers were involved (rising to 22,000 at its close), while the defenders numbered in excess of 40,000. That alone should have ensured a British victory. That the Battle of Crete ended in humiliating defeat was entirely due to the incompetence of the defenders: Freyberg, with his irrational preoccupation with a seaborne invasion; Puttick and Hargest, with their failure to order an immediate counter-attack at Maleme; Andrew, with his premature withdrawal from the airfield there; even the ordinary soldiers, who time and again proved no match for the determined paratroopers.

Kasserine Pass

On 14 February 1943, Axis forces launched a two-pronged offensive against American troops holding the mountain passes in central Tunisia. Within a week, they had driven the jittery Americans back to and through the Kasserine Pass, taking thousands of prisoners in the process. It was a stunning tactical victory that owed much to the woeful defence put up by the inexperienced Americans.

Four months earlier, during the nine-day Battle of El Alamein, the tide in North Africa had turned in favour of the Allies. Worn down by Montgomery's numerically superior Eighth Army (as the Western Desert Force had been renamed), Rommel's Italian-German panzer army began to retreat westwards along the coast. By mid-December, it had been pushed back to El Agheila in Libya.

Meanwhile, on 8 November, the British and Americans had launched Operation Torch – the landing of troops in Algeria to take the retreating Axis forces in the rear. But as they advanced on Tunis, the Germans and Italians rushed troops across the Mediterranean to its defence. Commanded by General Jürgen von Arnim, this new formation became known as the Fifth Panzer Army.

By Christmas 1942, a combination of inexperienced Allied troops, long supply lines, Axis command of the air, the superior strength of von Arnim's armour, and bad weather had stopped the Allies in their tracks. 'Gentlemen,' said General Eisenhower, at that time the Supreme Allied Commander in the theatre, to his senior officers, 'we have lost the race for Tunis.'

Tunisia is a country of two levels with a coastal plain to the east and mountains to the west. At the beginning of 1943, the Fifth Panzer Army was occupying the coastal plain, while the Allies were strung out along the Grande or Eastern Dorsal, the mountain range that runs from the northern coast to the salt marshes of the south. The only way of breaking into the Allied line was round its southern tip, at Gafsa, or by surging through the mountain passes at Pichon, Fondouk, Faïd and Maknassy.

The Allies were also astride the Petite or Western Dorsal, a smaller range of mountains that joins with the Eastern Dorsal in the north, but diverges further south so that the two ranges form an inverted 'V'. It has passes at Maktar, Sbiba, Kasserine and Feriana. In effect, the Allies had a double defence line of mountains; to reach the Allied rear, an attacker would have to penetrate passes on both the Grande and Petite Dorsals – a daunting task.

At the end of January 1943, Rommel's army crossed the Tunisian frontier and linked up with von Arnim's forces. The Axis line now ran right through the country, from north to south, with Rommel guarding its lower end. He also had troops facing Montgomery behind a chain of forts known as the Mareth Line. But Montgomery, hampered by his ever-lengthening supply lines, would not be in a position to attack for some time and Rommel was keen to take advantage of this. He therefore proposed to the Comando Supremo, the Italian High Command in Rome, a joint attack with von Arnim against the Allied forces in Tunisia. Once they had been defeated, the two Axis armies could then deal with Montgomery.

By now, the three Allied corps in Tunisia had been placed under the orders of Lieutenant-General Kenneth Anderson, commander of the British First Army. They were, from north to south, the British V Corps under General Allfrey, the

French XIX Corps under General Koeltz, and the US II Corps under Major-General Lloyd Fredenhall. Rommel's plan was to attack through the southern mountain passes towards the rear area of II Corps at Tebessa in Algeria.

On 28 January, the Comando Supremo approved Rommel's plan but put von Arnim in charge. With the 10th and 21st Panzer Divisions, the latter detached from Rommel's army, von Arnim was instructed to push through the Eastern Dorsal defiles of Pichon in the north and Faïd in the centre. He chose to concentrate his main attack at Pichon; a preliminary thrust at Faïd would only be a diversion.

Two days later, tanks and infantry of the 21st Panzer Division overwhelmed the 1,000 French defenders at Faïd. An attempt by two battalions of American infantry to retake the pass the following day ended in fiasco. Disorientated by the noise and smoke, men became lost and wandered in all directions. The confusion spread as Stuka dive-bombers and ME-109 fighters bombed and strafed the attackers, causing many to retire without orders. Meanwhile, the 10th Panzer Division had pierced the French defences at Pichon. But worried about over-extending his troops, von Arnim ordered both divisions on to the defensive.

These latest setbacks alarmed Anderson. Fearing that the French were too poorly equipped and the Americans too inexperienced to withstand a full-scale Axis attack, he decided to wheel back his forces in the south and centre to strengthen his position in the north. If attacked, he told Koeltz and Fredenhall, they were to withdraw their troops to the Western Dorsal. Fredenhall, however, took this to mean that such a move was inevitable and began to issue the necessary orders.

The Germans saw their opportunity. On 9 February, having flown to Tunisia from Rome, Field Marshal Albert Kesselring, C-in-C of the Axis forces in the Mediterranean and Italy, had a conference with von Arnim and Rommel at the Luftwaffe base at Rennouch. 'We are going to go all out for the destruction of the Americans [i.e. US II Corps],' Kesselring told them. 'They have pulled back most of their troops to Sbeitla and Kasserine . . . We must exploit the situation and strike fast.'

Rommel, he explained, would attack the Americans in the south, at the oasis town of Gafsa, while von Arnim would do the same further north at Sbeitla. Once through the mountain passes, the two German panzer divisions, supported by Italian armour, would drive north towards the Algerian port of Bône, thereby cutting the First Army's lines of supply.

'Meine Herren,' said Kesselring gravely, 'after Stalingrad, our nation is badly in need of a triumph.'

Von Arnim was enthusiastic, but pointed out that there was not enough fuel for such an ambitious scheme. However, they could inflict large losses on the French and Americans and force them to withdraw. If Kesselring agreed, he would attack on the 14th.

Rommel was happy to go along with this, despite his personal distaste for von Arnim, a poker-faced Prussian aristocrat – in fact, a baron – of the old school. 'What counts,' he said, 'isn't any ground we gain, but the damage we inflict on the enemy.'

Kesselring assented, telling Rommel later that if the attack was a success he would be placed in command of all the forces in Tunisia with the despised von Arnim as his subordinate. But this, in fact, was never Kesselring's intention, for he then spoke to Rommel's doctor, who recommended that his ailing chief should be sent back to Germany for a rest cure by the 20th. 'Let's give Rommel this one last chance of glory,' the Field-Marshal told a delighted von Arnim, 'before he gets out of Africa.'

By 13 February, the final plans had been decided. Von

Arnim's operation – codenamed *Frühlingswind* ('Spring Breeze') – was in effect a double pincer movement to capture the village of Sidi bou Zid and the hill of Djebel Lessouda, either side of the main road from Faïd to Sbeitla in the Western Dorsal. Under the overall command of General Ziegler, von Arnim's deputy, two battle groups from the 10th Panzer Division would sweep through the Faïd Pass and advance either side of Lessouda. Infantry would follow in their wake and attack Sidi bou Zid. At the same time, two battle groups from the 21st Panzer Division would go through the Maizila Pass, to the south, and approach Sidi bou Zid from two directions. More than 200 tanks, half-tracks and guns would be involved.

Rommel's operation – codenamed *Morgenluft* ('Morning Air') – was for a mixed detachment from the Afrika Korps and the Italian Centauro Division to attack the town of Gafsa. Its strength was about 160 tanks, half-tracks and guns, but they would be augmented by reinforcements from the 21st Panzer Division if von Arnim's attack went well. Consequently, Rommel's advance would start a day later.

Despite the fact that his intelligence sources predicted an attack through the passes, Fredenhall had done little to improve their defences. No minefields had been laid and no trenches dug. Furthermore, the troops of his three divisions – the 1st and 34th Infantry, and the 1st Armoured – were mixed up and placed in piecemeal positions. The reason: Fredenhall had not once left his underground command post, hidden in a remote canyon 70 miles behind the front line; all his troop dispositions had been made on the basis of maps of doubtful accuracy.

Holding the area due to be attacked by von Arnim's panzers was Combat Command A, the forward element of the US 1st Armoured Division, under Brigadier-General McQuillan (with his headquarters in Sidi bou Zid). Dug in on Djebel

Lessouda, about five miles to the north, was a composite group under Colonel Waters, comprising the 2nd Battalion of the 168th Infantry Regiment, a company of tanks, a platoon of tank destroyers (half-tracks towing 75mm guns), and a battery of self-propelled 105mm howitzers. During daylight hours, a force of tanks and infantry screened the hill by manning an outpost line on the plain near to the Faïd Pass.

According to a scheme devised by Fredenhall, Waters would hold up any Germans coming through the pass long enough for Colonel Hightower to counter-attack from Sidi bou Zid with a mobile armoured reserve of about fifty Sherman tanks and two artillery battalions. A similar arrangement was in place for the Djebel Ksaira, a hill guarding the southern approach to Sidi bou Zid from the Maizila Pass, held by the 3rd Battalion of the 168th Infantry, some engineers and several artillery pieces. Where Hightower was to launch his counter-attack if the Germans came through both passes at the same time had not been decided.

At Sbeitla, 20 miles to the west of Sidi bou Zid, was Combat Command C, the divisional reserve. Near Tebessa, 40 miles further on, was Fredenhall's headquarters and the corps reserve – a handful of artillery and tank-destroyer battalions and one of infantry.

The attack was opened at dawn on 14 February 1943 – St Valentine's Day – as two battle groups from the 10th Panzer Division rumbled through the Faïd Pass and sent the small covering force of American infantry and artillerymen reeling back in panic. To delay the Germans, Colonel Waters sent forward his fifteen Honey light tanks. But their 37mm shells made little impact on the thick armour of the German Tiger (which had only very recently come into service) and Mark IV tanks, and one by one they were knocked out. Before long, Waters's entire position on Djebel Lessouda had been cut off.

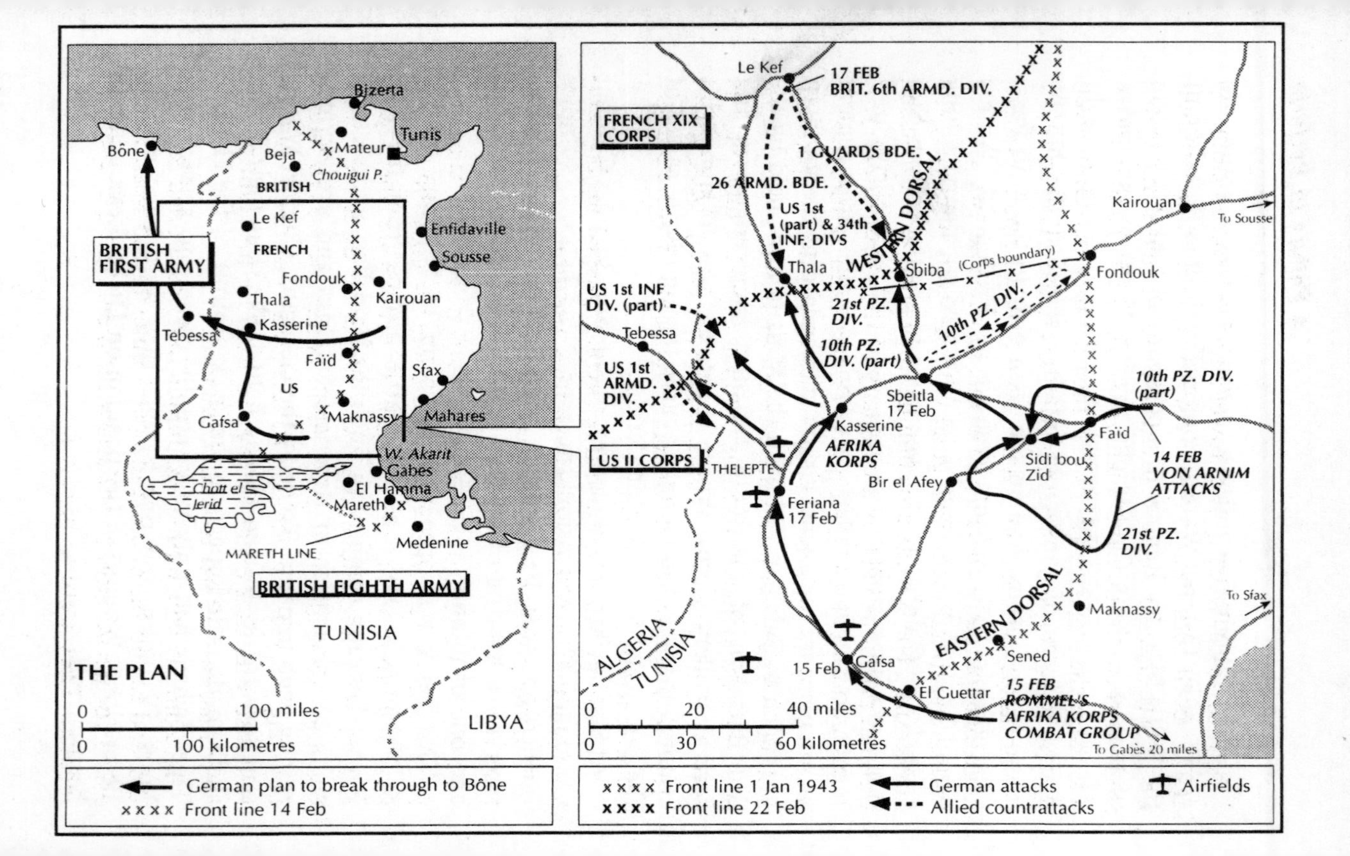

THE PLAN
Bizerta
Tunis
Mateur
Beja
Chouigui P.
Bône
BRITISH
BRITISH FIRST ARMY
Le Kef
FRENCH
Enfidaville
Sousse
Fondouk
Thala
Kairouan
Kasserine
Tebessa
Faïd
Sfax
US
Gafsa
Maknassy
Mahares
W. Akarit
Gabes
El Hamma
Mareth
Chott el Jerid
Medenine
MARETH LINE
BRITISH EIGHTH ARMY
TUNISIA
LIBYA
0 100 miles
0 100 kilometres
German plan to break through to Bône
x x x x Front line 14 Feb
Le Kef
17 FEB
BRIT. 6th ARMD. DIV.
FRENCH XIX CORPS
1 GUARDS BDE.
26 ARMD. BDE.
US 1st (part) & 34th INF. DIVS
WESTERN DORSAL
Kairouan
To Sousse
Thala
Sbiba
(Corps boundary)
Fondouk
US 1st INF DIV. (part)
21st PZ. DIV.
10th PZ. DIV.
Tebessa
10th PZ. DIV. (part)
US 1st ARMD. DIV.
Sbeitla 17 Feb
10th PZ. DIV. (part)
Kasserine
Faïd
AFRIKA KORPS
US II CORPS
THELEPTE
Sidi bou Zid
Bir el Afey
14 FEB VON ARNIM ATTACKS
Feriana 17 Feb
21st PZ. DIV.
To Sfax
EASTERN DORSAL
Maknassy
ALGERIA
TUNISIA
Sened
15 Feb
Gafsa
El Guettar
15 FEB ROMMEL'S AFRIKA KORPS COMBAT GROUP
0 20 40 miles
0 30 60 kilometres
To Gabès 20 miles
x x x x Front line 1 Jan 1943
x x x x Front line 22 Feb
German attacks
Allied countrattacks
Airfields

Sent to assist Waters, Hightower's mobile reserve came up against both 10th Panzer battle groups. The Shermans fought well, but they had a tendency to catch fire from even glancing hits (earning them the derisive nickname of 'Ronsons') and were no match for the German Tigers with their 88mm guns. Only seven out of the fifty-one Shermans would escape. The spineless performance of their supporting troops hardly helped: a reconnaissance company surrendered en masse, while the two artillery battalions abandoned their guns without even removing the firing pins. 'The Krauts are coming!' they cried as they rushed through positions held by a company of the 168th Infantry.

One or two officers, their pistols drawn, tried to intervene. It made no difference; wide-eyed with fear, they raced on. Colonel Drake, commanding the 168th Infantry Regiment, could see the débâcle from his command post 2 miles to the east of Sidi bou Zid. He immediately rang McQuillan. 'They're running away, General,' he said. 'Your men are running away.'

'You don't know what you're saying,' gasped an incredulous McQuillan. 'They're only shifting position.'

'Shifting position hell!' shouted Drake. 'I know panic when I see it.'

By now, Drake had problems of his own. Two battle groups from the 21st Panzer Division had come through the Maizila Pass and were about to encircle his troops on the Djebel Ksaira. At 11am, McQuillan rang Ward, his divisional commander, and told him that his whole command was in danger of being cut off. Fredenhall, in turn, was informed, but his response was that everyone was to hold tight. 'Continue on your mission,' Ward told McQuillan.

At 2pm, by which time McQuillan had abandoned Sidi bou Zid and set up his command post 5 miles to the west, Drake asked for permission to withdraw his men. Again Fredenhall refused. They were to hold on.

Next morning, Ward moved his reserve up from Sbeitla to counter-attack the Germans at Lessouda. Comprised of fifty-four tanks, a company of tank destroyers, two batteries of self-propelled artillery, and the 6th Armoured Infantry Regiment in trucks and half-tracks, it was a formidable force. But the tanks drove straight into an ambush and soon found themselves attacked by panzers from three sides. Artillery fire and Stukas added to the confusion.

At 6pm, the entire force was ordered to disengage. The infantry emerged relatively unscathed, as did the artillery, but the tank battalion had been annihilated. Only four tanks survived. In two days the 1st Armoured Division had lost ninety-eight tanks, fifty-seven half-tracks and twenty-nine artillery pieces either destroyed or captured.

Anderson had already ordered the evacuation of the garrison at Gafsa, enabling Rommel's forces to occupy it without a fight during the afternoon of the 15th. Now he told Fredenhall to withdraw all his forward troops to Sbeitla. Messages to this effect were eventually delivered by air to the two beleaguered battalions of the 168th Infantry on Lessouda and Ksaira. But it was far too late to save them. As well as 1,000 men of the 3/168th, Drake had about 600 miscellaneous troops under his command. Almost all were killed or captured as they attempted to reach safety. Of the 600 men of the 2/168th on Lessouda, only about half made it back.

On 16 February, Anderson instructed Koeltz to start withdrawing his right flank from the Eastern to the Western Dorsal. As a result, the US 34th Division, which was under French command, was moved to the Sbiba Gap, 20 miles north west of the 1st Armoured Division at Sbeitla.

During the afternoon, sixty panzers attacked the forward elements of the 1st Armoured Division – a tank battalion, two artillery batteries and an infantry regiment – at a crossroads 11 miles east of Sbeitla. Gradually, the Americans were pushed back to positions on the edge of the ancient town; for once, the withdrawal was not accompanied by the usual panic.

All this changed when the Germans launched an all-out assault by three columns of panzers that night. Within hours, the panicked defenders were melting away. 'Uncertain and nervous,' wrote Martin Blumenson, author of *Rommel's Last Victory*, 'fatigued and confused, hemmed in by widespread firing that seemed to be all around them, believing that the Germans were already in Sbeitla, demoralized by the piecemeal commitment and intermingling of small units . . . many men lost their confidence and self-discipline.'

The roads were soon jammed with vehicles heading to the rear. 'We just lost our heads,' admitted one soldier.

Under the impression that Ward's division was about to be overwhelmed – when in fact its front-line troops were providing stout resistance – Anderson gave permission for it to withdraw – but not before 11am the following day. This would give the French time to move back to Sbiba. When news arrived that Rommel was advancing from Gafsa towards the Feriana Pass, however, Anderson authorized a general withdrawal.

At dawn on 17 February, Fredenhall issued a series of orders. Ward's division would leave Sbeitla only when forced to do so and retire through the Kasserine Pass toward Thala. The withdrawal would be covered by a regiment of engineers who would pull back into the pass and organize its defence. If pressed, the composite force at Feriana, under Colonel Stark, was to move back to Tebessa. Once all these moves were completed, the new defensive line would face south, along a line from Tebessa eastward through Kasserine to Sbiba.

Unfortunately, events did not go to plan. When Ziegler's panzers attacked the south of Sbeitla in the early afternoon, a battalion of American tank destroyers held their ground for half an hour. They then withdrew to a rallying point but, finding it under fire, continued on into the town. There they joined a confused mass of transport heading towards Kasserine. Bombed by Stukas, many drivers lost control and ran their vehicles into ditches, while others simply abandoned theirs and fled on foot. Only a few elements of the division withdrew in an orderly manner. By 5pm, the town had been abandoned to the Germans. Further south, Rommel's troops had taken Feriana.

Operations *Frühlingswind* and *Morgenluft* had already exceeded all expectations. But instead of pressing on, von Arnim decided to dissolve Ziegler's command. Leaving the 21st Panzer Division at Sbeitla, he sent a battle group of tanks towards Sbiba and the rest of the 10th Panzer Division to a bivouac behind the Pichon and Fondouk Passes, where he still harboured ambitions.

Rommel, on the other hand, was all for exploiting the gains made so far. If all the available forces converged on Tebessa, the Allies would lose a major supply and communication centre. This, in turn, would make possible a strike deep to the rear that would force the Allies 'to pull back the bulk of their forces to Algeria, thus greatly delaying their offensive operations'. During the evening of the 17th, he telephoned von Arnim and and told him that his forces were too weak to take Tebessa alone. He needed the support of the 10th and 21st Panzer Divisions, and a holding attack further north to keep the bulk of the Allied forces busy. Von Arnim refused to help, claiming he did not have enough supplies for a deep penetration.

Next day, having failed to change von Arnim's mind, Rommel appealed to the Comando Supremo in Rome. He was counting on Mussolini's need for a 'victory to bolster up his internal political position'. It was almost midnight when a signal arrived from Rome authorizing Rommel to conduct the attack with both panzer divisions under his command. But instead of advancing north-westward through Tebessa, he was ordered to strike north towards Thala and Le Kef. In Rommel's opinion, this change was 'an appalling and incredible piece of shortsightedness' – for the direction of the attack was 'far too close to the front and bound to bring us up against strong enemy reserves'.

It was doubly unfortunate given the fact that General Sir Harold Alexander, Eisenhower's deputy, had ordered Anderson 'to concentrate his armour for the defence of Thala'. Alexander had wrongly assumed that Rommel was seeking a tactical victory rather than a strategic one; but thanks to the Comando Supremo, Anderson would have his troops in the right place when the Germans attacked. If Rommel had had his way, the Allies would have been in trouble, for most of the reinforcements rushed south were sent to Thala and the Sbiba sector east of it, while Tebessa was covered largely by the remnants of the US 1st Armoured Division.

Rommel's plan of attack, set for the morning of 19 February, was for his Afrika Korps combat group to lead the advance on Thala, and ultimately Le Kef, through the Kasserine Pass – midway between Sbeitla and Feriana. The 21st Panzer Division would also make for Thala via the Sbiba Gap. Since it would be unable to arrive in time to take part in the initial thrust, the 10th Panzer Division would be available to assist either breakthrough.

By 16 February, in line with Fredenhall's orders, Moore's 16th Engineer Regiment had laid almost 3,000 mines along both sides of the road between the village of Kasserine and the pass. Moore had then withdrawn his men through the pass and placed them in positions covering the exit. His engineers were on the right; an infantry battalion on the left. In all, 2,000 men were guarding a front of about three miles. Supporting them were two batteries of 105mm howitzers, a battery of horse-drawn French 'seventy-fives' (75mm field guns), and a battalion of tank destroyers. The intention was to destroy the Germans as they became bunched together in the pass.

But Moore's engineers, in particular, were of doubtful fighting ability. They were not trained to hold defensive positions and only one officer had combat experience. When a German reconnaissance unit probed the pass on the 18th, wounding one man, many engineers left their posts and fled to the rear. Some were rounded up but others could not be found. With the inexperience of the engineers in mind, Fredenhall put Colonel Stark of the 26th Infantry Regiment in overall command.

The German attack at Kasserine began at 7.30am on 19 February with a series of probes by the veteran Afrika Korps infantry. Meeting stiff resistance at the end of the pass, many began to climb the heights. Then tanks were brought up to assist the infantry, but a combined attempt to rush the pass was broken up by artillery and machine-gun fire.

'Bring on the Krauts!'

On St Valentine's Day 1943, when the Germans launched their two-pronged offensive in the mountains of eastern Tunisia, the Allied cause was not helped by the fact that the US II Corps, the formation upon which the hammer blow would fall, was commanded by an officer as laughable as Major-General Lloyd Fredenhall.

He had been appointed by General 'Ike' Eisenhower, the Supreme Allied Commander, on the strength of his 'brilliant leadership' during the landings at Oran the previous November. But their success had been due more to a lack of serious resistance from the Vichy French troops than to any bravura performance on the part of the Americans. Fredenhall, of course, thought differently: his men had fought brilliantly; bring on the

Krauts! This arrogant amateur was soon to be rudely awakened.

On 13 February, the day before the Germans attacked, Eisenhower went to visit his corps commander. He was shocked to discover that not once had Fredenhall left his command post, situated in a remote canyon more than 70 miles behind the front line. Fearful of discovery by enemy aircraft, he had ordered his engineers to spend three weeks digging deep air-raid shelters. His only contact with his forward units was by telephone; all troop dispositions had been made using large-scale maps spread out on the floor of his command post.

Once the battle was under way, Fredenhall's orders were often so obscure as to be indecipherable. 'Move your command,' read one extraordinary message, 'that is, the walking boys, pop guns, Baker's outfit and the outfit which is the reverse of Baker's outfit and the big fellow to M, which is due north of where you are now, as soon as possible. Have your boss report to the French gentleman whose name begins with J at a place which begins with D which is five grid squares to the left of M.'

Hardly surprising, then, that inexperienced troops led by such a commander should have suffered the sequence of humiliating defeats culminating in the loss of Kasserine Pass. The line was eventually stabilized by the arrival of British armour, but Fredenhall's number was up. His replacement, George S. Patton, had all of his swagger, and more, but none of his caution. Patton led from the front.

All the while, Stark was gaining in confidence, particularly when reinforcements arrived in the shape of a battalion of the 39th Infantry Regiment and some British mortars and reconnaissance troops. He was now considerably stronger than the

Afrika Korps combat group attacking him (three weak battalions – one panzer and two infantry).

To strengthen his line, Stark placed one company of the 39th on either flank. But the Germans continued to infiltrate along the heights and behind the defensive positions, at dusk capturing 100 men from the company guarding the engineers' flank. This, and the sudden withdrawal of their supporting tank destroyers, caused the neighbouring company of engineers to abandon their positions. The Germans promptly occupied them.

Meanwhile, the 21st Panzer Division's assault at Sbiba had been foiled by the strong Allied force there – eleven infantry battalions to the attackers' two, and a superiority in tanks and guns (the 21st Panzer had less than forty serviceable panzers during the operation). Believing that the 'Allies were weaker at Kasserine than at Sbiba', Rommel decided to focus the weight of his attack there by throwing in the 10th Panzer Division the following day. Unfortunately, von Arnim had kept back almost half of it, including the battalion of Tiger tanks, and Rommel was forced to make do with one tank battalion, one motorcycle battalion, and two panzer-grenadier (motorized infantry) battalions.

The attack grew in intensity from midday onwards. Eventually, panzers and *Nebelwerfers* (multiple rocket launchers) were brought forward, and all five infantry battalions were thrown in. The engineers were the first to crack and by 5pm the whole defensive line had disintegrated. Hundreds surrendered; the rest scarpered to the rear. The only effective resistance was provided by a detachment from the British 6th Armoured Division. Made up of a tank squadron, an infantry company and a field battery, it was covering the road behind the pass. The end came when the veteran 8th Panzer Regiment drove the British squadron's eleven tanks back against the mountains and destroyed them, the survivors abandoning their vehicles and fleeing on foot over the hills. In all, some twenty tanks and thirty armoured troop-carriers (most towing 75mm anti-tank guns) were captured, causing Rommel to marvel at how 'fantastically well equipped' the Americans were.

That night, Rommel sent reconnaissance groups northwards along the Kasserine–Thala road and westwards towards Tebessa – but they made no contact with the enemy. Expecting a counter-attack nevertheless, he spent the next morning on the defensive. When the enemy failed to appear, he pushed on up the road to Thala with the remnants of the 10th Panzer Division – thirty tanks, twenty self-propelled guns and two panzer-grenadier battalions.

About ten miles on from Kasserine, having already overrun a British anti-tank company, they came up against the rest of the 26 Armoured Brigade group, consisting of two armoured regiments and two infantry battalions. Once again the panzers prevailed, forcing the British to withdraw. Attempts were made to hold successive ridges, but each time the British were outflanked and enfiladed by their numerically inferior foe. At dusk, as they withdew into prepared positions at Thala, Rommel's armour was still hard on their heels. Led by a captured Valentine tank, the German panzers were mistaken for stragglers and allowed to enter the British lines. There they caused panic and confusion, destroying vehicles and capturing 700 soldiers before fresh Allied units forced them back. For the loss of twelve tanks, they had knocked out nearly forty.

Thala was the furthest point of the German advance. With air reconnaissance indicating the imminent arrival of powerful Allied reinforcements, and the Afrika Korps' thrust towards Tebessa having been checked by elements of the American 1st Armoured and 1st Infantry Divisions, Rommel realized that no further gains could be made and decided to break off the engagement. Montgomery's Eighth Army, fast approaching from the east, would have to be dealt with first.

In just eight days, the Germans had advanced up to 80 miles, taken more than 4,000 prisoners and destroyed nearly 200 tanks. Their own casualties were little more than 1,000, with fewer than 50 tanks lost. For attacking troops, such ratios of losses are unprecedented and were largely the result of the shameful lack of resistance offered by the inexperienced Allies – particularly the Americans.

Afterword

In 1992, intoxicated by the recent collapse of the Soviet Union, the American political scientist Francis Fukuyama wrote in *The End of History and the Last Man:*

> What we may be witnessing is not just the end of the Cold War, or the passing of a particular period of post-war history, but the end of history as such: that is, the end point of mankind's ideological evolution and the universalization of Western liberal democracy as the final form of human government.

Fukuyama was right in one sense. There was not a single liberal democracy with universal suffrage in 1900; a century later the number had risen to 119 (or 60 per cent of all forms of government). Such liberal democracies, it seems, are less likely to go to war with each other than other political systems.

Yet Fukuyama was predicting nothing less than the end of war between nation states. This was a little optimistic. In recent years we have seen the unprovoked invasion of Iraq by a Western coalition (chiefly US and British forces) and Russia's war in Georgia. The reemergence of authoritarian Russia and China as economic and military superpowers is another reason to doubt that Fukuyama is right.

Meanwhile the various forms of low-intensity warfare – civil wars, insurgencies and terrorism – have shown no signs of abating and, chiefly thanks to the rise of radical Islam (notably al-Qaeda and its many affiliations), NATO forces have seen more action during the last 20 years than at any time since the Vietnam War. 'Everything changed with the end of the Cold War,' wrote General Sir Peter Inge, Britain's Chief of Defence Staff, in 1996. 'Up to then we knew what we had to do and we knew who the potential enemy was and we could train and prepare accordingly. Now it's all guesswork.'

NATO, as a result, has fought in recent years (or is still fighting) in places as diverse as Somalia, Bosnia, Kosovo, Sierra Leone, Iraq, Afghanistan and, most recently, Libya. Israeli, in response to Hezbollah attacks, has invaded Lebanon. Inevitably, blunders have been made in all these wars. The most notable are as follows:

Black Hawk Down

In early October 1993, in Mogadishu, Somalia, a botched US special forces' mission to capture two senior militia commanders resulted in a two-day running battle that cost the American military over 90 casualties and the loss of two Black Hawk helicopters. So shaken was the American President Bill Clinton that he ordered the withdrawal of all US troops from Somalia, followed a year later by the remaining UN contingents.

The US troops had arrived in Somalia, on the horn of Africa, in December 1992 as part of a UN mission to separate warring militias, distribute food and save lives. So successful were their early efforts that all Somali political groups agreed at a Conference on National Reconciliation, in March 1993, to work towards peace and democracy. Yet it soon became clear that the most powerful faction, the Habr Gidr clan led by General Mohamed Farah Aidid, was not prepared to implement the agreement. Unwilling to share power with the smaller factions, Aidid began to broadcast anti-UN propaganda on Radio Mogadishu. The response of the local UN commander, Lieutenant-General Cevik Bir, was to order the closure of the radio station. As a precursor to this, a Pakistani-led force was sent to the station on 6 June to seize arms. But it was attacked en route by Aidid's militia, and lost 24 killed and 61 wounded (including three Americans). The following day the UN Security Council passed Resolution 837, effectively declaring war on Aidid and his men.

Over the next few months, various botched efforts were made to capture or assassinate Aidid, including one helicopter attack on a compound in Mogadishu that killed between 20 and 50 people. Four western journalists were killed in revenge by angry mobs. In late August, with casualties mounting on both sides, President Clinton sent in a special task force of 450 US Army Rangers and Delta Force commandos to apprehend Aidid and his lieutenants.

Known as Task Force Ranger, this new elite band was commanded by Major-General William F. Garrison, a tough no-nonsense Vietnam veteran and former commander of Delta Force (the US Army's equivalent of the SAS). On 3 October, Garrison received intelligence on the whereabouts of two of Aideed's senior lieutenants, Omar Salad and Abdi Hassan Awale. He at once ordered their capture in an operation that was code-named Gothic Serpent.

The plan was for Delta Force commandos and four chalks of Rangers to be flown in Little Bird and Black Hawk helicopters from US Headquarters at the airport, south of Mogadishu, to the target building in the heart of the city. While the commandos landed and assaulted the building, the Rangers would abseil (or fast-rope) from the Black Hawks and set up a four-corner defensive perimeter. The capture of Salad and Awale was timed to coincide with the arrival of a ground convoy of nine Humvees and three five-ton trucks, under the command of Lieutenant-Colonel Danny McKnight, which would be used to transport the entire assault team and their prisoners back to base. Including the ground force, and the pilots of the helicopters and spy planes (tasked with filming the operation), there were 160 men involved. The entire operation was expected to take no more than an hour if everything went to plan. Inevitably it did not.

The first setback was when a Ranger missed his grip on the fast rope and fell 70 feet to the ground, badly injuring his head and neck. He required immediate evacuation. Meanwhile thousands of Somalis were pouring onto the streets, many armed, as word spread that an American operation was in progress. 'Kasoobaxa guryaha oo iska celsa cadowga!' ('Come out and defend your homes!') shouted Aidid militiamen into megaphones. Somalis responded by erecting barricades across the main roads and lighting tires to summon help.

By the time the ground convoy arrived at the rendezvous a few minutes late, having taken a couple of wrong turns, the operation was almost over. The Ranger chalks were under fire but were holding their own at the blocking points; while the Delta commandos had captured not two but three of Aidid's senior lieutenants, the original targets and an unexpected bonus, Abdi Yusuf Herse. It was 3.50 p.m. and, apart from the injured Ranger, everything had gone like clockwork.

The plan was about to unravel. Three humvees were detailed to evacuate the injured Ranger back to the airport. No sooner had they set off than they were forced to run a gauntlet of machine-gun fire, rocket-propelled grenades, burning tires and barricades. The first fatality was an M-60 gunner, shot in the head. Others were wounded before the tiny convoy reached the safety of the airport perimeter.

Back at the target building, meanwhile, a lack of communication between the assault force and the ground convoy had meant a loss of vital minutes as both assumed the other was not ready to leave. Once this was sorted out, Delta commandos were preparing to move the captured Aidid lieutenants into one of the 5-ton lorries when an explosion was heard overhead. A Black Hawk helicopter, providing covering fire over the target area, had been hit in the tail by a rocket-propelled grenade. Within seconds it had plummeted to the ground, killing both pilots and severely injuring the two crew chiefs. As two of the survivors, both Delta Force snipers, emerged from the wreckage they were fired at by enraged Somalis.

Watching the drama unfold in real time, using a feed from cameras on high-level observation helicopters, were General Garrison and his staff at the airport. They knew at once that the initiative had been lost, and that a closely planned snatch operation had become an unscripted rescue mission. The entire assault and ground force would have to fight its way through to the crash site, a few hundred yards to the northeast of the target building, before Aidid's forces were able to cut it off and kill all the survivors.

When the first rescuers – crew from a Little Bird and half a chalk of Rangers – reached the crash site they discovered that the two Delta Force snipers needed emergency evacuation: one had damaged his arm in the crash and the other had been shot four times holding off the militia. They were both loaded onto the Little Bird, but the more serious casualty, 25-year-old Staff Sergeant Daniel Busch, later died of his wounds. He was posthumously awarded the Silver Star for gallantry in the face of the enemy.

As more rescuers – including a helicopter-borne Combat Search and Rescue Team (CSAR) and the remaining nine vehicles of the ground convoy – raced towards the crash site, a second orbiting Black Hawk was hit. It went down to the south of target building. A second rescue operation was needed, and a further division of the tiny 150-strong American ground force; and all the time resistance was stiffening and casualties mounting.

The order was for the ground convoy to meet the assault team at the first crash site, gather up the survivors and then head for the second downed Black Hawk. But first the convoy lost its way, losing vital time; and then so severe was the Somali cross-fire, and so heavy the convoy's casualties – with more than half its seventy-five occupants hit by bullets or shrapnel – that its commander chose to abort his mission and head back to the airport.

Meanwhile the rest of the assault team of Delta Force commandos and Rangers had reached the first crash site, having taken a number of casualties en route. But once there they could do no more than reinforce those already there in a string of buildings close to the first crash site. There they would remain until the following morning, beating off repeated attempts by Aidid's men to overrun them with small-arms fire and rockets from Little Bird helicopters (crewed by men of the 160th Special Operations Aviation Regiment, the only air unit trained to fight at night).

The survivors of the second crash site, therefore, were largely left to fend for themselves. A quick reaction force was sent to their aid from the US base at the airport; but facing stiff opposition it would arrive far too late to intervene. Instead the only assistance the crash survivors received was from two Delta Force snipers, Sergeants Randy Shughart and Gary Gordon, who insisted on being dropped from a separate Black Hawk. They found both pilots alive, but with broken legs. So having lifted them out of the crashed helicopter, they set up a defensive position to wait for the quick reaction force. It never arrived.

For a time the two Delta Force commandos managed to hold off the approaching Somali mob, inflicting numerous casualties. But eventually Gordon was shot and then, 15 minutes later, so too was Shughart. One of the injured pilots, Mike Durant, managed to shoot two assailants before he was captured and badly beaten, his nose broken and his eye-socket shattered. Durant was spared execution by the intervention of a militia commander who hoped to trade the American pilot for Somali captives.

The remaining 99 soldiers of the assault and CSAR teams at the first crash site – many of them wounded – were finally relieved in the early hours of 4 October by a huge UN convoy of more than 100 vehicles. But there was only enough room in the vehicles for the casualties, and many unwounded Rangers and Delta Force commandos had to cover the first mile of the withdrawal on foot.

Total US casualties were 18 killed and 73 wounded; a further eight Malaysians and two Pakistanis were wounded (one of the Malaysians fatally). Somali casualties were far more numerous, with at least 500 militiamen and civilians killed and up to a thousand wounded. Though badly mauled, the US forces had not been defeated. But the shocking sight of dead US soldiers – Shughart, Gordon and a crew chief from the second downed Black Hawk – being dragged through the streets of Mogadishu by cheering Somali mobs was enough to prompt President Clinton to order all American troops to leave Somalia by March 1994. The only consolation was the release of injured pilot Mike Durant after eleven days of captivity.

Thanks to a combination of factors – but chiefly poor planning and an underestimation of the enemy – the US troops in Somalia had won a tactical victory but suffered a strategic defeat. Without American muscle, there was no way the UN could force General Aidid to agree to a political settlement. Soon after the departure of all UN forces in spring 1995, Aidid declared himself president of Somalia; but his regime was never internationally recognized, nor could it end the fighting. Aidid was mortally wounded in a gun battle with former militia allies in July 1996. Somalia is still at war.

Operation Certain Death

In late August 2000, in the tiny west African state of Sierra Leone, 11 soldiers from the 1st Royal Irish Regiment (1 RIR) and a liaison officer from the Sierra Leonean Army were captured by a drug-crazed rebel force known as The West Side Boys. The operation to rescue them was one of the most audacious by British Special Forces since World War Two. Yet it risked, for a time, becoming the British version of Black Hawk Down; and if the RIR soldiers had not taken an unscheduled trip into bandit country, it would not have been necessary in the first place.

The soldiers were part of 200-strong contingent of 1 RIR that had been in Sierra Leone for just four weeks. They had followed in the wake of a fully-armed British intervention force – paratroops, special forces, combat aircraft, attack helicopters and warships – that had reached the country in May 2000 to bolster a UN peacekeeping force (UNAMSIL), and to prevent the capital Freetown from being overrun by a vicious rebel force known as the Revolutionary United Front (RUF) whose chief aim was to gain control of the country's diamond fields.

The initial intervention – codenamed Operation Palliser – stopped the rebel advance in its tracks and restored a measure of order to Freetown. Thereafter British personnel took up key posts in the government, central bank and police, and began the task of rebuilding the country's army so that it could crush the remaining rebels. The 1 RIR contingent was part of this latter effort.

Shortly after 11.30 a.m. on 25 August, a three-vehicle patrol set off from 1 RIR's tented camp on an old colonial plantation to the south-east of Freetown. It contained two officers, five NCOs and an interpreter-guide from the Sierra Leonean Army (SLA) called Corporal Mousa Bangura. The destination of the patrol was the town of Masiaka, 35 miles to the north-east of Freetown, where a Jordanian battalion of UN peacekeeping troops had its headquarters. The RIR had never patrolled that far before, and their commander, Major Alan Marshall, was keen to extend his unit's area of operations. But to get to Masiaka they would have to pass through the Occra Hills, territory controlled by a violent, unpredictable and trigger happy rebel group known as the West Side Boys (or the West Side Niggas, as they styled themselves).

The journey to Masiaka passed without incident, though the tattooed and bullet-festooned rebels manning the West Side Boys' checkpoints seemed high on cannabis and very jumpy. A third of the way back, at a UN checkpoint, Major Marshall was told about a left turning ahead that led, via a dirt track, through jungle to the village of Magbeni on the banks of the Rokel Creek. En route were several small settlements that were suffering from food shortages and a lack of medicines. It was an area, as far as the Jordanians knew, that was not under rebel control, and one that would hugely benefit from the arrival of aid agencies. Would the British patrol be prepared to check it out? Marshall was asked.

Marshall pondered the request. He knew that to leave the tarmac road was risky, and that Magbeni was in the heartland of the West Side Boys' territory. But he was swayed by the dual prospect of opening up the area to aid and, at the same time, gathering intelligence on the West Side Boys' strength and activities. His biggest error, however, was in failing to consult Corporal Mousa before he made his decision. If he had he done so, he would have learnt that Magbeni was the West Side Boys' hidden jungle headquarters.

Twice Corporal Mousa, travelling in the third of the three Land Rovers, voiced his concerns as they drove deeper into bandit country. 'I'm not really comfortable with this, sa,' he said to a Captain Flaherty. 'The West Side Boys are very unpredictable. I don't think we should be going into their area.'

'Come on, Corporal,' replied Flaherty, 'you have to admit you are feeling afraid now, aren't you?'

Mousa tried a third time, insisting that they stop and ask Major Marshall why there were 'going to this place'. But before Flaherty could answer they emerged from the jungle into a clearing filled with ramshackle buildings. They had reached Magbeni. Within seconds the vehicles had been surrounded by a swarm of heavily armed rebels and brought to a halt. 'This is de West Side Niggas' area!' shouted a rebel machine-gunner, blocking their way. 'What de Fuck you doing in dis our area?' On either side of him stood fighters – some as young as nine – armed with Rocket-propelled grenade launchers (RPGs), self-loading rifles (SLRs) and AK47 assault rifles, all aimed at the stationary Land Rovers.

As Major Marshall got out of the lead vehicle to speak to the rebels, a lorry mounted with a heavy machine-gun cut off the patrol's line of retreat. If the British had been following standard procedure, there would have been a hundred yards between each vehicle and enough time to react. Instead they had been driving far too close and were all trapped.

Some of the British soldiers wanted to shoot their way out. But they were persuaded to put down their weapons by their superiors who knew there was little prospect of escape. Then all were robbed and beaten by the drug-crazed rebels, though none as savagely as Mousa who, in a 30 minute ordeal, was knocked unconscious.

For almost a week, as negotiators tried to secure their release, the eleven RIR captives were forced to endure mock executions and regular beatings. But at least they were locked in a building, unlike the badly injured Mousa who was kept in a waterlogged pit in the ground that he later described as 'hell on earth'. On the sixth day the rebels agreed to free five of the RIR captives in return for a satellite telephone and medical supplies. Major Marshall had previously agreed that the most junior ranks would leave first. But now he changed his mind and chose the five married men: Captain Flaherty and four NCOs. The four young Rangers, two still in their teens, felt betrayed. But two were granted a reprieve when Flaherty, the signals officer, and his corporal were told they had to stay to assist negotiations.

The remaining seven captives – two officers, three NCOs (including Mousa) and two Rangers – were finally rescued in a brilliantly audacious joint SAS/SBS operation, supported by men of the 1st Battalion, The Parachute Regiment (1 Para), on 10 September. The operation – codenamed Barras, or Certain Death to those who took part – was authorized by Tony Blair, the Prime Minister, but only after negotiations had broken down and the rebels had threatened either to kill the hostages or move them to a separate location.

One SAS trooper, Brad Tinnion, was killed in the assault and 20 British soldiers were wounded (one seriously). Rebel losses were at least 150 killed (including women and adolescent boys) and 18 taken prisoner, among them the rebel leader Foday Kollay. Blair later paid tribute to the 'skill, professionalism and courage' of the assaulting troops, and added: 'There really are no finer armed forces anywhere in the world.'

The bravery awards included two Conspicuous Gallantry Crosses (second only to the Victoria Cross), four Military Crosses, Five Distinguished Flying Crosses (for the helicopter pilots involved) and a number of Mentions in Despatches (MDs). But the assault had come closer to disaster than Blair's eulogy and the list of medals would imply. 'Someone was being very overconfident thinking they would just leg it when they heard the fire,' said a Para who took part. 'You have got to ask yourself why we took so many hits. Taking one in ten casualties is fairly serious. It was, as they say, a damn close run-thing. It was just that far from being a total fuck-up.'

Inevitably Major Marshall took the lion's share of the blame for allowing his small but heavily armed patrol to be captured by, according to some reports, an initial rebel force of just 25 fighters. Marshall 'made an error of professional judgment', stated an MOD report of the incident, 'in diverting from a planned and authorized journey so as to make an unauthorized visit to the village of Magbeni. There his patrol was overwhelmed. Maj. Marshall made a grave mistake.'

But by reminding Brigadier Gordon Hughes, the commander of British forces in Sierra Leone, that the country remained 'an unstable and volatile environment and that the deployment of his forces was to be strictly controlled,' the MOD report hinted at other failings up the chain of command. The whole British contingent, it seems, had badly underestimated the threat posed by a small and unruly, but battle-hardened, rebel group.

The Jamiat

In September 2005, at Basra in southern Iraq, a failed SAS surveillance operation sparked a chain of events that strained Anglo-US military relations to the limit and resulted in a permanent breach between British troops and the Iraqi police they were supposed to be training. It was a blow from which the British counter-insurgency effort in Iraq would not recover.

The target of the SAS surveillance operation was a Captain Jafar of the Iraqi police who was suspected of abusing his power as head of Basra's Serious Crimes Unit to extort money and settle personal feuds. The plan – authorized by British Army commanders in Basra – was for two SAS men to follow Jafar back to his house so that an arrest could be organized away from the public gaze. The alternative was to pick him up at his place of work, Jamiat police station in the heart of the old city; but army commanders decided not to take that option because it could have led to a firefight with the Iraqi police and the exposure of the uncomfortable truth that many in the Iraqi security forces sympathized with local Iranian-backed Shiite militias like Muqtada al-Sadr's Jaish al Mehdi.

According to a report by US freelance journalist Steve Vincent, printed in the *New York Times* on 25 July 2005, police officers were 'perpetrating many of the hundreds of assassinations' in Basra, while 'the British stand above the growing turmoil, refusing to challenge the Islamists' claim on the hearts and minds of police officers'. Vincent quoted an Iraqi police lieutenant: 'The British know what's happening but they are asleep, pretending they can simply establish security and leave behind democracy.' Two days after his article was published, Vincent was apprehended on the street by men in police uniforms, beaten and eventually shot. He became the first US journalist to be killed covering the insurgency.

Just over two months later, in the early morning of 19 September, two SAS men had just finished a preliminary surveillance of Captain Jafar, and were driving in their battered car towards one of the main thoroughfares in east Basra, when they were stopped at an Iraqi police checkpoint. Ordered to get out of the car, they refused and a firefight broke out. At least one policeman was killed before the SAS men drove off. But when their underpowered banger – designed to blend in with the locals – was overhauled by police cars, they surrendered. 'They wanted to try and talk their way out of it,' was how one SAS colleague put it.

The men were taken to Jamiat police station – a fact confirmed by a second surveillance team – where they were stripped, beaten and filmed. As soon as word spread of their capture, the SAS commander in Iraq flew a team of 50 men – 20 members of A Squadron, a platoon of Paras from the recently deployed Special Forces Support Group, four signalers and a medic – from Baghdad to Basra to assist with their rescue.

Meanwhile the response of the British commander in Basra, Brigadier John Lorimer, was to tread carefully. He had been advised by his superiors at the Permanent Joint Headquarters (PJHQ) in north London not to take any action that might inflame the situation. So, instead of an assault force, he sent a negotiating team and a cordon of British soldiers to block off the main routes in and out of the Jamiat. This prompted a riot by infuriated locals, and a sergeant of the Staffordshire Regiment was badly burned when a Molotov cocktail hit his Warrior armoured vehicle.

It was now that the Iraqi Interior Ministry responded to British diplomatic pressure by ordering the men's release; but Jafar and his colleagues at the Jamiat refused to obey. Convinced that their colleagues' lives were in imminent danger, the SAS in Iraq and Hereford pressed hard for permission to launch a rescue attempt. But the Chief of Joint Operations could not be contacted – it was later suggested that he was playing golf and had switched off his mobile phone – and nobody else had the authority to approve such a mission.

Eventually, after hours of procrastination, Brigadier Lorimer ordered an attack on the Jamiat to be spearheaded by Challenger tanks and warriors. Before it could take place, however, the captives were smuggled through the cordon to a separate location nearby. It was fortunate for them that this move was spotted by a British surveillance helicopter, and the news relayed to the SAS team that had arrived from Baghdad and was making its final preparations on the outskirts of the city. Its commander altered his plan accordingly, and allocated just a couple of SAS troopers to accompany the army attack on the Jamiat – which would serve as a useful distraction – while the rest of the assault team stormed the new location. The second operation does not appear to have been authorized, even by Lorimer. Yet it was necessary, say the SAS, because they knew from intercepts that their men had been handed to a radical fringe group called Iraqi Hezbollah and were in danger of summary execution.

The attacks went in at 9 p.m., with Lorimer's armour crushing cars and a couple of temporary buildings as it burst into the Jamiat compound. At the same time the SAS team blew in the doors and windows of a nearby house and found their comrades in a locked room; the gaolers had all fled, probably warned by dickers (lookouts) that an attack was imminent. The troops involved were jubilant: the assaults had rescued the men with no casualties. The backlash, however, was swift.

The first to protest was the Governor of Basra Province, Mohammed al-Waeli, who described the assault on the Jamiat as 'barbarous' and ordered his police to end all cooperation with the British. In the weeks to come, British soldiers would be turned away and sometimes threatened when they arrived at Iraqi police stations to mentor and supervise. The rottenness at the core of Britain's effort to ensure security for ordinary Iraqis had been exposed.

And it was not just Iraqi officials who were angry. So, too, were the Americans – Britain's Coalition partners – who had not been consulted during the crisis; and the SAS who, convinced their government had responded weakly to the crisis, were about to 'down their tools' and refuse to operate in Iraq when the rescue attempt was sanctioned. 'The incident brought out a huge number of issues,' recalled one member. 'The infiltration of the IPS [Iraqi Police Service] by the Iranian

Revolutionary Guards, and the lack of will on the UK's part to name but two.'

There was a suspicion among the American military – correct, as it turned out – that the British were, by their inaction, actually making things worse in Basra; that they were so determined to leave the country as soon as possible that they were ignoring the realities on the ground in southern Iraq. Comments by senior British officers seemed to bear this out. Soon after he took over as Chief of General Staff (head of the British Army) in the summer of 2006, General Sir Richard Dannatt told a reporter that he hoped we would 'get ourselves out [of Iraq] sometime soon because our presence exacerbates the security problems'.

In December 2007, judging that the Iraqi security forces were capable of looking after themselves, British troops were withdrawn from Basra and consolidated at the airport. It was left to American troops – buoyed by the success of their own recent 'surge' in central Iraq – to support an Iraqi-led operation to clear unruly militias – particularly Muqtada al-Sadr's Mahdi Army – from much of Basra in a spring 2008 offensive known as the 'Charge of the Knights'. By the end of the year, the city was as peaceful as at any time since the invasion – but little credit was due to the British Army. Its troops formally withdrew from operational responsibilities in Iraq on 30 April 2009.

The '33-day War'

On 12 July 2006, near the village of Zarit on the Israeli-Lebanon border, a small group of well-trained Hezbollah fighters ambushed an unsuspecting patrol of Israeli Defence Force (IDF) soldiers in two high-visibility vehicles. Three Israelis were killed, two wounded and two seized by Hezbollah and dragged back across the frontier, vanishing deep into Lebanon.

They were the first shots in a 33-day war that lasted until a UN-brokered ceasefire was enforced on 14 August, by which time Israel had suffered a resounding defeat. It had 'failed', according to Ron Tira, the IDF campaign planner, 'on the strategic, operational and tactical levels'. And the reasons were many: overconfidence, because of Israel's counterinsurgency success in Palestine; a lack of proper training for IDF troops in conventional warfare; and an overreliance by the Israeli commander-in-chief on airpower.

Israeli shortcomings, moreover, were magnified by the excellence of the Hezbollah fighters. The Iranian-backed terrorist organization had been preparing for conflict with Israel since the IDF's ignominious withdrawal from south Lebanon in 2000, aiming to exploit what its leader Hasan Nasrallah believed to be Israel's Achilles heel: 'Israeli society, [which] is as weak as a spider's web.'

Its small cadre of fighters had been thoroughly trained in complex tactics and modern weaponry, much of it supplied by Syria and Iran. They had constructed extensive underground networks of tunnels, bunkers and communication links; they had devised a long-term intelligence scheme for infiltrating the Israeli secret services and turning enemy agents; and a cunning deception plan had been fabricated to create overconfidence in the IDF command and conceal the transformation of Hezbollah from the limited guerilla outfit of the 2000 conflict to a new and sophisticated semi-conventional fighting force.

Israel's errors had begun long before 12 July 2006. The success of its recent operations in Palestine had given rise to a false and overconfident assumption that it was too powerful to be attacked by local enemies. Israel's political and military leaders believed they did not have to prepare for a 'real war' in Lebanon since any future challenge would be swatted aside by the IDF's superior firepower.

As a consequence, army training was reduced to a minimum during the years 2000 to 2006 as defence budgets were slashed. According to one reserve armoured battalion commander, tank crews needed at least one 'five-day refresher exercise' a year to be in sufficient state of combat readiness. Yet 'most hardly got that in the course of three years, others in the space of five, and yet others none at all'. Infantry reservists, making up the bulk of the IDF's fighting force, had received no manoeuvre training since 2000 because they were too busy with occupation duty in the Palestine. Even the IDF's regular forces had not completed a major manoeuvre operation for more than a year by the summer of 2006.

Ignoring the threat of Hezbollah in the Lebanon – or indeed the possibility of a major war against a coalition of surrounding Arab countries – the IDF had concentrated instead on the challenge posed by the Second Palestinian Intifadah, which had broken out in 2000. From that date, according to one Israeli historian, 'the IDF went on a massive retooling [effort] to be an urban anti-terrorism force, like a large SWAT team and became the most advanced large scale anti-terrorism force in the world'.

But this preoccupation with urban fighting in the occupied territories of Palestine meant that the IDF would find itself trained and equipped for a very different type of combat to that the one it encountered in the Lebanon in 2006. By then the IDF was expert at 'conducting cordon and search operations, door-to-door searches, hasty raids, and identifying and capturing or killing suspected Palestinian terrorists and guerillas'. It had gained, in addition, extensive intelligence on Palestinian terror organizations like Hamas, and its control of the borders of Gaza and the West Bank meant that its opponents were restricted to outdated weaponry.

Hezbollah, on the other hand, had not rested on its laurels

after victory in the 17-year struggle with Israel in 2000. Instead, anticipating a future Israeli incursion into southern Lebanon, it had used the Viet Cong as inspiration for an extensive tunnel system across the IDF's likely routes of approach. It had also built up stockpiles of weapons – particularly short and medium range rockets and anti-tank guided missiles – and constructed steel-reinforced, camouflaged bunkers throughout its area of operations. Unlike in the occupied territories, Israeli intelligence in Lebanon signally failed to identify these skilful preparations for a future conflict.

Other failings included 'missing, obsolete or broken equipment' for the IDF's reserve forces, as identified by the government-appointed Winograd report into the war. But the single greatest failing was Israel's military leadership. Many senior officers had not trained with their troops for years and were unfit for their posts, men like the reserve armoured division commander Erez Zuckerman, a former Marine commando, who had no experience of tanks or mechanized forces. There was, stated the Winograd report, 'deficient Israeli preparedness'.

No one exemplified this deficiency more than the IDF's new Chief of the General Staff, Lieutenant-General Dan Halutz. An airman and the first member of the Israeli Air Force (IAF) to command the IDF, Halutz was convinced – like 'Bomber' Harris before him – that planes alone could win a war. 'We . . . have to part with the concept of a land battle,' he had declared in 2001. Instead he espoused an American military strategy known as Effects Based Operations that required precision firepower to knock out an opponents 'cognitive domain' – including command centres, logistical hubs, radars and transport links – and little or no land forces' activity. Halutz, however, had misunderstood the concept of EBO. When first used by US troops in Iraq in 2001, the intention of EBO had been to disrupt the enemy and prepare the ground for land troops to exploit. It was not a substitute for land warfare, as Halutz believed. The consequences of this misreading of EBO were disastrous for Israel when war flared in the summer of 2006.

Wednesday 12 July was the final day of guard duty for the reserve troops at milepost sector 105 on the Israel-Lebanon border. Overconfident and underprepared, the seven Israeli reservists began their ill-fated patrol without the standard and mandatory briefing, inspection or dispatch order from their commanding officer. Demob happy, they were carrying their personal civilian effects in their vehicles and did not even bother to dismount from their vehicles as they approached milepost 105.

They paid the price when hidden Hezbollah fighters detonated an improvised explosive device (IED) under one of the vehicles and destroyed the other with anti-tank missiles. As two of the survivors were dragged across the border, Hezbollah fired rockets and mortars across the whole sector to slow the Israeli response. It worked. More than half an hour had elapsed before 'Hannibal', the pre-planned kidnap rescue mission, was authorized. It then took a further nine minutes for helicopters to reach the ambush site; an hour and a half for armoured vehicles to begin the pursuit; and two and a half hours for 'Fourth Dimension', artillery attacks on 69 bridges in southern Lebanon, to be launched. All were ineffective, and the pursuit was soon called off.

It was now up to Israel's politicians – Prime Minister Ehud Olmert and Defence Minister Amir Peretz – to decide how to respond to Hezbollah's act of aggression and they did so by declaring war on Hezbollah. The Israeli strategy, devised chiefly by Halutz, was to rely on precision airstrikes to destroy Hezbollah's central nervous system so that its limbs – its frontline fighters – could not function. Halutz's aim was to avoid the use of land troops entirely. Yet when his planes could only destroy an estimated 7% of Hezbollah's military power, he was forced to launch a ground attack with an army that was largely made up of poorly trained reservists.

The first serious assault was against the south Lebanese town of Bint Jbeil, a site of huge symbolic significance because it was from here that Hasan Nasrallah had delivered his stirring victory speech in 2000, including the line: 'Here, too, blood has broken the chains that bind us, and has humiliated despots and arrogant men.' Part of Halutz's strategy was to create a 'consciousness of victory' in Israel and a 'cognitive perception of defeat' among the leaders of Hezbollah, hence the choice of Bint Jbeil as his first major objective. If it fell, as he assumed it must, he would have struck a crucial psychological blow from which Hezbollah might not recover.

Thus, on 24 July 2006, troops from the IDF's 7th Armoured Brigade and Golani Infantry Brigade moved into positions on the north east of the town. That night two Merkava IV tanks were destroyed, and two IDF soldiers killed, by concealed IEDs. The news caused Halutz to order Lieutenant-General Udi Adam, commander of the IDF Northern Forces, to take the town with a single battalion of tanks. Adam's protests were in vain, and at dawn the following day two armoured companies tried to enter the east of the town. They were met by a storm of anti-tank missiles, mortars, rockets, machine-gun and sniper fire from rooftops, bunkers and concealed trenches. Many tanks and a third of the IDF soldiers were hit, with the rest forced to withdraw, a humiliating defeat that caused the Israeli media to accuse the IDF of 'idiotic military maneuvers' and to question the competency of senior commanders. Bint Jbeil remained in Hezbollah hands for the rest of the war, a propaganda disaster from which Halutz and the IDF did not recover.

A second major setback took place on 12 August, near Wadi al-Saluki, as the UN-brokered ceasefire was about to take effect. (A day earlier, the UN Security Council had unanimously agreed Resolution 1701, which called for a halt to the fighting and the deployment of 15,000 peacekeepers to help the Lebanese Army take control of southern Lebanon.) The aim of the Israeli attack by 24 Merkava IV tanks of the 401st Armoured Brigade was to capture the town of Ghadouriyeh a vital strategic position on the Litani River. To allow the tanks to advance safely, paratoopers of the Nahal Infantry Brigade were first dropped on the outskirts of Ghadouriyeh and Farun, an adjacent town, to secure the approach route across the wadi, a tributary of the Litani and in an area covered with dense undergrowth that confined all vehicles to the partially built road. But they simply occupied buildings in the two towns before confirming – incorrectly – that the area had been secured.

Convinced it was entering 'safe' territory, the lead tank of the Israeli armoured column had just crossed a bridge over the wadi when Hezbollah fighters detonated a mine under it, killing all its occupants. The rest of the column, trapped on the road, was then engaged with a swarm of rockets and anti-tank missiles. It was, recalled one Israeli tankman, 'like being shot at in a shooting range'. Another added: 'The first missile which hits is not really the dangerous missile. The ones which come afterwards are the dangerous ones – and there always follow four or five after the first.' Incredibly, not one of the 24 trapped tanks used its smoke grenades to make it harder for the enemy to hit them. By the battle's end the following morning, 11 of the 24 tanks had been hit by either missiles or RPGs, with eight crewmen killed.

The verdict on the operation by UN spokesperson Timur Goksel, who visited the terrain, was withering: 'Anyone dumb enough to push a tank column through Wadi Saluki should not be an armoured brigade commander but a cook.' The 401st Armoured Brigade could have bypassed the wadi either to the south or on the more northern road to Farun; its failure to do so gifted Hezbollah yet another propaganda victory on the last day of the war.

Since 5 August the IDF had had more than 30,000 troops in southern Lebanon; their opponents just 10,000. Yet the IDF failed even to prevent Hezbollah from targeting northern Israel with more than 100 rockets a day, and 250 in the hours before the ceasefire came into effect on 14 August. A chastened Ehud Olmert admitted that mistakes had been made, but the Israeli press was unimpressed. 'You cannot lead an entire nation to war promising victory,' opined the *Ha'aretz* newspaper, 'produce humiliating defeat, and remain in power. You cannot bury 120 Israelis in cemeteries, keep a million Israelis in shelters for a month, and then say "Oops, I made a mistake."'

But survive he did, despite the Winograd Report's censure that the Prime Minister, Defence Minister and Chief of Staff had all made 'very serious failings' in their conduct of the war. Olmert, in particular, had made 'mistaken and hasty judgements and did not manage events, but [rather] was dragged along' by ill-advised options proposed by General Halutz.

Overconfidence, lack of preparation, inadequate training, poor political and military leadership, defective strategy and tactics, lack of communication between various arms of the military – all had contributed to the IDF's humiliation. But the quality of their opponents should also be taken into account, and due credit given to the excellence of the Hezbollah leaders and rank and file who, unusually, had not rested on their laurels of their 2000 success.

Further Reading

GENERAL

Brian Bond, *Fallen Stars* (Brassey's, 1991); Norman F. Dixon, *On the Psychology of Military Incompetence* (Futura, 1988); Charles Messenger, *Great Military Disasters* (New York, 1991); James Perry, *Arrogant Armies: Great Military Disasters and the Generals Behind Them* (New York, 1996); Geoffrey Regan, *The Guinness Book of Military Blunders* (Guinness, 1991); ibid., *The Guinness Book of More Military Blunders* (Guinness, 1993).

1 Unfit to Command

The Retreat from Kabul: Peter Hopkirk, *The Great Game* (Oxford, 1991); Patrick McCrory, *Signal Catastrophe* (The History Book Club, 1967).

The Charge of the Light Brigade: John Sweetman, *Balaclava 1854* (Osprey, 1990); Donald Thomas, *Charge! Hurrah! Hurrah!* (Omega, 1976).

Antietam: John Priest, *Antietam: The Soldier's Battle* (Oxford University Press, 1993); Norman S. Stevens, *Antietam* (Osprey, 1994).

Spion Kop: Thomas Packenham, *The Boer War* (Weidenfeld & Nicolson, 1980); Oliver Ransford, *The Battle of Spion Kop* (John Murray, 1969).

Suvla Bay: Lyn Macdonald, *1915: The Death of Innocence* (Headline, 1993); Nigel Steel & Peter Hart, *Defeat at Gallipoli* (Macmillan, 1994).

The Fall of Singapore: Timothy Hall, *The Fall of Singapore* (Methuen, 1983); B.H. Liddell Hart, *History of the Second World War* (Cassell, 1970).

2 Planning for Trouble

The Jameson Raid: Hugh Hole, *The Jameson Raid* (Philip Allan, 1930); Jean van der Poel, *The Jameson Raid* (Oxford University Press, 1951).

Colenso: Rayne Kruger, *Good-bye Dolly Gray* (Cassell, 1959); Ian Knight, *Colenso 1899* (Osprey, 1995).

The First Day on the Somme: Martin Middlebrook, *The First Day on the Somme* (Allen Lane, 1971); Lyn Macdonald, *Somme* (Michael Joseph, 1983).

The Dieppe Raid: John Mellor, *Forgotten Heroes* (Methuen, 1975); Brian Loring Villa, *Unauthorized Action* (Oxford University Press, 1989).

Arnhem: Cornelius Ryan, *A Bridge Too Far* (Hamilton, 1974); John Baynes, *Urquhart of Arnhem* (Brassey's, 1993).

Bravo Two Zero: Andy McNab, *Bravo Two Zero* (Bantam, 1993); Chris Ryan, *The One That Got Away* (Century, 1995).

3 Meddling Ministers

Bannockburn: William Mackenzie, *The Battle of Bannockburn* (Glasgow, 1913); Ronald McNair Scott, *Robert the Bruce* (Hutchinson, 1982).

Sedan: John Bierman, *Napoleon III and his Carnival Empire* (New York, 1988); Richard Holmes, *The Road to Sedan* (Royal Hist. Soc., 1984).

St Valéry: Saul David, *Churchill's Sacrifice of the Highland Division* (Brassey's, 1994); Ernest Reoch, *The St Valery Story* (privately printed, Inverness, 1965).

North Africa 1940–1: Christopher Hibbert, *Benito Mussolini* (Penguin, 1965); Roger Parkinson, *The War in the Desert* (Book Club Associates, 1976).

Stalingrad: John Erickson, *The Road to Stalingrad* (Weidenfeld & Nicolson, 1983); Edwin Hoyt & Edwin Palmer, *199 Days* (Robson, 1993).

Goose Green: Mark Adkin, *Goose Green* (Leo Cooper, 1992); Max Hastings & Simon Jenkins, *The Battle for the Falklands* (Pan, 1983).

4 Misplaced Confidence

Teutoburger Wald: J. F .C. Fuller, *The Decisive Battles of the Western World: Volume I, 480* BC–1757, (Eyre & Spottiswoode, 1954).

The Second Crusade: Steven Runciman, *A History of the Crusades:*

Volume II, The Kingdom of Jerusalem (Cambridge University Press, 1952)

Custer's Last Stand: Evan S. Connell, *The Son of the Morning Star* (Pavilion, 1985); E. Lisle Reedstrom, *Custer's 7th Cavalry* (Cassell, 1992).

Isandhlwana: Rupert Furneaux, *The Zulu War* (Weidenfeld & Nicolson, 1963); Donald R. Morris, *The Washing of the Spears* (Simon & Schuster, 1965).

Yalu River: Max Hastings, *The Korean War* (Michael Joseph, 1987); Michael Schaller, *Douglas MacArthur* (Oxford University Press, 1989).

Dien Bien Phu: Bernard Fall, *Hell in a Very Small Place* (Pall Mall Press, 1967); Pierre Langlais, *Dien Bien Phu* (Paris, 1963).

5 A Failure to Perform

Crécy: Jean Froissart, *Chronicles* (Penguin, 1978); Barbara W. Tuchman, *A Distant Mirror* (Macmillan, 1979).

Caporetto: Cyril Falls, *Caporetto 1917* (Weidenfeld & Nicolson, 1966); Ronald Seth, *Caporetto* (Macdonald, 1965).

The Kaiser's Battle: Robert Asprey, *The German High Command at War* (Little, Brown, 1993); Martin Middlebrook, *The Kaiser's Battle* (Allen Lane, 1978).

Anual: Rupert Furneaux, *Abdel Krim* (Secker & Warburg, 1967); Martin Woolman, *Rebels in the Rif* (Oxford University Press, 1969).

Crete: Anthony Beevor, *Crete: The Battle and the Resistance* (Penguin, 1992); Ian Stewart, *The Struggle for Crete* (Oxford University Press, 1955).

Kasserine Pass: Martin Blumenson, *Rommel's Last Victory* (Allen & Unwin, 1968); Rutherford Ward, *Kasserine* (Macdonald, 1971).

Afterword

Gilbert Achcar, *The 33-Day War* (London, 2007).

Mark Bowden, *Black Hawk Down* (London, 1999).

Benjamin S. Lambeth, *Air Operations in Israel's War Against Hezbollah* (Santa Monica, 2011).

Damien Lewis, *Operation Certain Death* (London, 2004).

Nicholas Noe (ed.), *Voice of Hezbollah: The Statements of Sayyed Hassan Nasrallah* (London, 2007).

Mark Urban, *Task Force Black: The Explosive True Story of the SAS and the Secret War in Iraq* (2010).

Index